The Best of
Annals of Improbable Research

AIR

Annals of Improbable Research

CAUTION:

Highly Reactive Mix

Active Ingredients:

Science, Technology, Medicine, Literature, Art.
Also contains 3% biodegradable filler material in accordance
with academic minimum daily requirements.

Recommended Dosage:

One issue every two months. Can be taken with meals.
Additional dosage of Internet *mini-AIR* can be taken monthly.

WARNING:

Contents are unexpectedly educational and informative,
especially in patients who suffer allergic reactions to science,
technology, literature, or art. Can be highly addictive.

The Best of
Annals of
Improbable Research

edited by

Marc Abrahams

W. H. Freeman and Company
New York

Interior design by Diana Blume
Cover and interior illustrations by Roy Wiemann

Library of Congress Cataloging-in-Publication Data

The best of annals of improbable research / edited by
    Marc Abrahams.
        p.   cm.
    ISBN 0-7167-3094-4
    1. Science.   2. Humor.   3. Annals of improbable
research.   I. Abrahams, Marc.   II. Annals of improbable
research.
Q172.B47   1997
502′.07—dc21
                                        97-14025
                                        CIP

Printed in the United States of America

First printing, 1997

Contents

Items marked with a star (*) are based on material taken straight from standard research (and other Official and Therefore Always Correct) literature. Many of the other items are genuine, too, but we don't know which ones.

Dedication and Special Thanks vii

CHAPTER **1**
The Improbable History of *AIR* 1

From Irreproducible to Improbable 6

Kinetics of Inactivation of Glassware 8

CHAPTER **2**
Nobel Thoughts 15

James Watson 16

Roald Hoffmann 17

Dudley Herschbach 19

Richard Roberts 21

Mel Schwartz 23

David Baltimore 24

Linus Pauling 25

William Lipscomb 27

Sidney Altman 29

CHAPTER **3**
Ig, Ig, Ig Nobel—
A Different Kind of Prize 31

We *Are* Amused* 34

The Ig Nobel Prize Winners 35

Those Who Covet the Ig 45

Transmission of Gonorrhoea
Through an Inflatable Doll* 46

Of Mites and Man* 47

Failure of Electric Shock Treatment for
Rattlesnake Envenomation* 49

The Okamura Fossil Laboratory* 51

CHAPTER **4**
Astronomy, Physics and Food 59

Chaos: Evidence for the Butterfly Effect 60

The Ubiquitous Holy Grail* 63

Sleep Research Update 66

Scheduled UFO Sightings 66

A Curious Particle Accelerator
in Switzerland 67

The Laser Cheese Raclette 68

Nanotechnology and the
Physical Limits of Toastability 72

The Aerodynamics of Potato Chips 75

The Effects of Peanut Butter on the
Rotation of the Earth 79

Mondocentrism 81

The Correlation Between Tornadoes
and Trailer Homes 82

Low Probability of Any Further
Abduction by Aliens* 85

Scientific Gossip 87

May We Recommend* 88

AIR Vents 89

CHAPTER **5**
The New Chemistry 91

Apples and Oranges: A Comparison 93

Xerox Enlargement Microscopy (XEM) 95

Ask Symmetra 97

Science Demonstration: Scratch 'n' Smell 98

Quantum Interpretation of the
Intelligence Quotient (QI of IQ) 99

The Politically Correct Periodic Table   101
Cindy Crawford Discovers*   102
AIR Vents   104

CHAPTER 6
Biology and Medicine   105

The Taxonomy of Barney   107
The Sad Crab of South Africa   113
The Ability of Woodchucks
    to Chuck Cellulose Fibers   114
The *mickeymouse* Gene   116
A Natural History of the
    Articulated Lorry   117
Cyclic Variation in Grass Growth   120
Arivederci, Aroma: An Analysis
    of the New DNA Cologne   121
Happy Yeast   123
A Man, a Woman, a Yeast*   124
Nematodes and Hieroglyphs   125
The Surfer Girl Fungus   127
Scientific Dining: Blackford Hall   128
Scientific Gossip   130
May We Recommend*   132
AIR Vents   134

CHAPTER 7
Medicine and Biology   135

How Dead is a Doornail?   137
Mystery Pheromone Coupon   139
A *glioblastoma multiforme* that
    Resembles Little Bird   140
Fifty Ways to Love Your Liver   141
The Medical Effects of Kissing Boo-Boos   142
Fetal Man in the Moon   145
The Guide to Politically Correct
    Cardiology   146
The Pop-Up Medical Thermometer   148
The Tomb of the Unknown Dentist   149
The Dental Micro-Luger   150
Scientific Gossip   151
May We Recommend*   152
Boys Will Be Boys*   155
AIR Vents   157

CHAPTER 8
Math and Models   159

Advances in Artificial Intelligence   161
The Mathematics of Telephone
    Numbers   162
The Value of Love, Using the
    Bob Dylan Model   164
The Paradigm Paradox   166
Mathematics: An Anagrammatical,
    if Pointless, Tale   167
SymmetraCal   168
Annual Swimsuit Extravaganza   169
The Studmuffins of Science
    Calendar Project   170
Scientific Gossip   171
AIR Vents   172

CHAPTER 9
Education, Scientific and Otherwise   173

Teachers' Guide   174
The Dead in the Classroom   175
A Mechanism for Getting and
    Keeping Students' Attention   177
Gummy Worm on a Sidewalk   178
*AIR*head Science Limericks   179
Virtual Academia: Year 1 Report   182
Scientific Gossip   183
AIR Vents   184

CHAPTER 10
Irrepressible Research   185

How to Write a Scientific Paper   187
Furniture Airbags   189
Internet Barbie and the Time Caplet   190
Internet Adventures   193
Project *AIR*head 2000*   199
With God in Mind   201

AIR info   202

Subscription Form   203

Index   205

Dedication

Alexander Kohn died a few weeks before the first issue of *AIR* appeared. In all his endeavors, Alex was a wise, kind, funny champion of curiosity and common sense, and a fearless, insightful agent provocateur against the spread of jargon, pap, and self-deception. By day, Alex was professor emeritus of virology at Tel Aviv Medical School. By evening, he was a scholar of the history of science. (His book *False Prophets* is a wonderful explanation and history of scientific fraud. Another book, *Fortune or Failure,* explores the role that serendipity plays in scientific research.) At night, often in the guise of Dr. X. Perry Mental (a guise that he and Harry Lipkin shared!), Alex produced some of the funniest and most literate commentaries, parodies, and satires ever written. Alex was loved and admired by many people in many places. Anyone who has read his work, and especially those of us who were lucky enough to know him personally, miss him terribly.

This book is dedicated to Alex, and to my parents, and to my sister, Jane, and my improbable, irrepressible niece and nephew, Kate and Jesse.

Special Thanks

AIR would not exist without the help of many extraordinary people. Some are mentioned in the following pages. Let me single out a few very special people here. They deserve extra emphasis and generous helpings of chocolate. Sid Abrahams, Stanley Eigen, Mark Dionne, Sip Siperstein, Nicki Sorel, Jerry Lettvin, Bob Rose, Amy Gorin, Dudley Herschbach, Bill Lipscomb, Rich Roberts, Shelly Glashow, Bob (Smitty) Smith, Deb (Symmetra) Kreuze, Howard Zaharoff, Karen Hopkin, Lynn and Steve Baum, Len Finegold, Lois Malone, Miriam Bloom, Jim Stoll, Jim Mahoney, Brenda Twersky, Steve Nadis, Jo Rita Jordan, Roland Sharrillo, Jon Connor, Chris Small, Jerry Lotto, Ariane Cherbuliez, Gary Dryfoos, Joe Wrinn, and that ever-productive, ever-elusive pair, Stephen Drew and Alice Shirrell Kaswell, have all gone to astounding lengths many times and in various ways both to create wonders and to prevent disasters. If you ever find yourself either in a jam or in need of a usefully brilliant idea, choose any two of them and do whatever it takes to enlist their aid and company.

Good agents and good editors are rare creatures, and must be treasured. My agent Regula Noetzle has proved to be as surprising and wonderful and dependable as her name. Holly Hodder, my editor at W. H. Freeman and Company, is a generous, reliable source of good ideas, clear (and accurate and practical!) criticism, and perfect doses of encouragement. Thanks also to the other folks at W. H. Freeman who worked so hard to make this book: Kate Ahr, Diana Blume, Patrick Farace, Paul Rohloff, Sheridan Sellers, and Susan Wein.

And to Martin Gardner, who pointed me toward the path of irreproducibility and improbability:

$$\sum_{i=1}^{\infty} i \text{ thank you.}$$

CHAPTER 1

The Improbable History of *AIR*

The *Annals of Improbable Research,* also known as *AIR,* is many things. First, *AIR* is a science humor journal. Now, hearing that, you might be tempted to toss this book aside, because maybe:

a) you don't like science and won't understand the book; or

b) you love science and know that science is too important to let people laugh at it.

Either way, you might be right. But I doubt it.

You don't like science? I'll bet that you've never seen how very human and quirky and charming and downright enjoyable science is to the people who spend their lives doing it. Yes, scientists and doctors and science teachers are *people,* not inhuman geniuses. Not most of them, anyway. You won't understand the book? I'll bet you will. Don't let yourself be intimidated by that spectacularly bad science teacher you had in seventh grade. Science is not about memorizing stuffy words and useless facts. Science is about asking questions—the "dumber" and simpler the better—and impishly, persistently trying to get sensible answers.

Science is too important to let people laugh at it? Ha. Science is too human, too much fun, and too important *not* to laugh at. An *AIR* editorial board member once approached the distinguished but ever-somber astronomer Carl Sagan and suggested that Sagan join our little gang of mischief-makers. Sagan, as the story was passed to me, replied tartly that what we are doing is "dangerous because it causes people to laugh at scientists." I think Sagan misunderstood us. What we're about is getting people to laugh *with* scientists as they laugh at this crazy universe and at themselves. For a more eloquent take on the general idea, look at the splendid editorial on page 34, "We *Are* Amused," which is reprinted from the British science journal *Chemistry & Industry.* It concerns Sir Robert May, the chief science advisor to the British government. May, too, had expressed his displeasure with our activities.

1

AIR 1:1 (January/February 1995). This was the premier issue.

Not Just for Scientists

AIR is also a humor, not just science humor, magazine. Scientists often tell us that *AIR* is the only journal they subscribe to that their families and friends also read. *AIR* is both cheaper and, dare I say it, more upbeat, more wide-ranging, and even more interesting than *Gastroenterology Motility Studies*, *IEEE Transactions on Dielectrics and Electrical Insulation*, or *Powder Technology*.

AIR covers virtually every subject imaginable, though often from a viewpoint that has never before been imagined. Our readers and our writers come from all walks, runs, and crawls of life—scientists, doctors, engineers, technicians, journalists, librarians, lawyers, moviemakers, students, English teachers, Swedish teachers, Hebrew teachers, German teachers, Chinese teachers (you get the idea... lots of

teachers), football players (both types), baseball players, synchronized swimmers (though they tend to read the *Annals* asynchronously), artists, plumbers, roofers, ministers, rabbis, priests, nuns, and auctioneers.

Thinking Is Possible, Even in School

AIR is also, believe it or not, a teaching tool, and not just at the college and postgraduate levels. Middle and high school teachers like to copy *AIR* articles and simply give them, without comment, to students. Within three minutes some student will pipe up, "Hey, wait a minute. What is this?" At that point begins what good teachers love and bad teachers dread: the students spend the rest of the day, and maybe the rest of the week, asking all kinds of questions. Curiosity has arrived on the scene. Take a look at the "Teacher's Guide" on page 174 for tips on how to start this off, and at the letter from a proud parent on page 184 about the unexpected, wonderful effect on a 14-year-old of the article called "A Natural History of the Articulated Lorry." (The Lorry article itself is on page 117.)

We at *AIR* have a serious intent involving more than just formal education. In our benevolently megalomaniacal way, we are trying to seduce people everywhere into the habit of *thinking about what they are told* by television, magazine and newspaper reports and by Official Persons. Even in the most revered newspapers and television news reports, many of the stories on any subject, science-related or not, are fully as absurd, or as sensible, as anything you will find in *AIR*. Any claim that comes from an Official Source is worth at least a moment's thought before you decide to accept it.

Stranger, To Be Sure, Than Fiction

AIR is also, and proudly, a journal of record for many of the world's most colorful and stupefyingly impressive activities. Generally, about half the contents of each issue are straightforward reports about genuine research, culled from the more than 10,000 "serious" research journals being published these days. Our readers see us as the central clearinghouse for their favorite research reports, and send us a constant stream of photocopies, faxes, and e-mail. We indicate which items are genuine (see the table of contents), and usually give enough information that you can go to the library and look up the original research report. Each issue of the *Annals* is crammed with several collections ("*AIR*-head Research Review," "*AIR*head Medical Review," "*AIR*head Legal Review," etc.) of such stuff. Some gems appear in this book in the sections called "May We Recommend" and "Boys Will be Boys."

How does *AIR* compare with the other science journals? A reader named Avraham Sonenthal answered that question in a letter to the editor:

> You describe *AIR* as being "the journal of inflated research and personalities. . . ." You're in good company. That could apply to just about every scientific journal in existence.

True, true, but there's a very important point I want to emphasize. A piece of research can be both (a) funny and (b) good science. And it can also be (c) important. On the other hand, it can be just (a).

Here, There, Everywhere

AIR comes out six times a year, and goes to readers in many countries. We have a large presence on the Internet, too. Our free monthly newsletter, *mini-AIR*, is crammed with tidbits too tiny or too timely to make it into the magazine. You can get details by sending e-mail to our automatic information address <info@improb.com> and by looking at our web site, the appropriately named Hot*AIR* (which is at http://www.improb.com). You can reach us by mail at: *AIR*, P.O. Box 380853, Cambridge, MA, USA. We also show up here and there on radio and TV (especially on ABC News's improbable wee-hours-of-the-morning news program *World News Now*). From time to time you can find us in the most unexpected places presenting lectures and slide shows about improbable research and the Ig Nobel Prizes. The American Association for the Advancement of Science for some reason has us present a special improbable research seminar at its annual meetings.

Ig, Ig, Ig Nobel

And then there are the Igs. In 1991, with help from quite a few friends and colleagues, I started a little ceremony to honor people whose achievements "cannot and should not be reproduced." We handed

AIR 2:1 (January/February 1996) was a special issue devoted to the Fifth First Annual Ig Nobel Prize Ceremony. The cover featured two of the five Nobel Laureates who recited the poem "DNA and Green Eggs and Ham."

out ten Prizes. Actually, the Prizes were handed out by four genuine Nobel (as opposed to Ig Nobel) Laureates wearing Groucho glasses. Since then, the little ceremony has grown to an annual sellout event held in the largest auditorium at Harvard University, attended by 1200 magnificent eccentrics, and broadcast over National Public Radio's *Talk of the Nation/Science Friday*, the C-SPAN television network, and the Internet. Last year one of the winners, Dr. Harold Moi, co-author of the medical report "Transmission of Gonorrhoea Through an Inflatable Doll," flew to Cambridge from his home in Oslo, Norway *at his own expense* to accept the Ig Nobel Public Health Prize (and four other winners accepted their Prizes either in person or through representatives). From his seat on stage at Sanders Theater, Dr. Moi had a splendid view of the world premiere and only performance of "Lament Del Cockroach," a mini-operetta for mezzo-sopranos and Nobel Laureates. You can learn more about the

Igs by looking in Chapter 3, "Ig, Ig, Ig Nobel—A Different Kind of Prize." The Seventh First Annual Ig Nobel Prize Ceremony will take place in October 1997. I hope that henceforth you will keep your eyes peeled for individuals worthy of receiving an Ig Nobel Prize. Send us your nominations, please.

Our Curious *Irreproducible* History

We are a curious bunch, by any definition of "curious." And we have a long history.

In 1955, Alexander Kohn, a virologist by trade, concocted a research paper called "Kinetics of Inactivation of Glassware." It described the various ways that beakers, test tubes, and the like vanish from the laboratory. At the top of this report, Alex wrote the heading, "*Journal of Irreproducible Results*, Volume 2, Number 1." Soon thereafter he teamed up with Harry Lipkin, a physicist, and together they edited and published the *Journal* for many years from their base in Israel. The *Journal* grew and thrived, to the point where handling the subscriptions (people actually wanted to pay for such a thing!) was driving Alex and Harry to distraction. Eventually Alex and Harry made arrangements with an outside party that would handle the subscriptions, leaving them free to oversee the contents. Eventually, this arrangement brought problems, which I will not go into here but which Harry will be delighted to tell you about, should you buy him a cup of coffee. Harry has prepared a concise history of the *Journal*. It appears on page 6.

Many years passed and the *Journal* withered considerably from its glory days. It was in 1990 that I wandered onto the scene.

I had been writing all sorts of things for years and foisting them on tolerant friends, but had never really tried to get any of it published. I had been spending my time in the software world, working on such things as the Kurzweil Reading Machine for the Blind, and then starting a company called Wisdom Simulators, where we simulated complex experiences from a variety of professions. Eventually, I sent some of my articles to Martin Gardner, whose wise, funny, math/science/literary column in *Scientific American* I had enjoyed immensely until that sad day in the early 80's when Martin retired. In case you never ran across Martin's column, I strongly suggest that you get yourself to a bookstore and treat yourself to one of his books.

Martin kindly encouraged me to track down an address for something called the *Journal of Irreproducible Results*, for which he had occasionally written but which, he thought, had probably gone out of business. Still, he said, you might as well try, since there's no other place to get science humor published.

So I went to the library, dug up an old address for the *Journal*, and sent off my package of articles. As an experiment, I sent an identical package to *The New Yorker*, addressed to "Fiction Editor." *The New Yorker*'s fiction editor then sent back a note saying that I had made a mistake, that I should re-send the articles to the "Nonfiction Editor." At about the same time, I got a phone call from the publisher of the *Journal of Irreproducible Results*. He asked if I would like to be the *Journal*'s editor. A few days later, after having actually seen a copy of the *Journal*, I became its editor.

One of the first things I did was write to Alex Kohn, who was happy that someone new was going to try to bring the *Journal* back to life. Alex became my mentor, largely via mail and later e-mail, and occasionally by telephone. In the Fall of 1991, Alex and I spent the better part of a week together at a conference at Woods Hole, where much was revealed to me. Harry Lipkin, too, got back into the swing of things, and with help from both founders and from everyone else I could talk into it, the *Journal* began growing rapidly. I initiated a series of interviews with Nobel Laureates, several of whom turned out to be pranksters of the highest order and were quite eager to become part of the quickly growing gang. And in 1991, with help from a lot of wonderful and wonderfully eccentric people, I started the Ig Nobel Prize Ceremony, which almost immediately became known as "The Ig." As mentioned above, Chapter 3 tells the story in some detail.

From *Irreproducible* to *Improbable*

Improbably and unpredictably, the company that owned the *Journal* had a shuffling of management, and suddenly became not at all keen on seeing the *Journal* continue. For several years, we irreproducible folk tried to change their minds, and even tried to purchase the *Journal* and publish it ourselves. Eventually, though, it became very, very clear that there was no way we could save the *Journal* as Alex and Harry and I and the rest of the forty years' worth of editorial staff had known it.

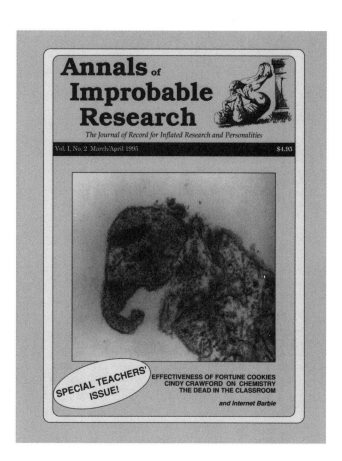

AIR 1:2 (March/April 1995) was a special teachers' issue. It also introduced Internet Barbie to an unsuspecting world.

And so, in early 1994 we walked away from the *Journal* and started everything anew. Thus was born the *Annals of Improbable Research*. Alex Kohn chose the new name, as he had devised the old one.

Having no resources, and not even having the old *Journal* subscribers list, we began life on the Internet, in stripped-down fashion. A year or two earlier, I had begun writing a newsletter and distributing it via e-mail. So in 1994 I started a new newsletter, *mini-AIR*. *mini-AIR* grew rapidly in readership and, in various ways, in influence. Eventually it carried the announcement that we were accepting subscriptions for the soon-to-appear print magazine the *Annals of Improbable Research*. Many people, from many places, sent in checks. Many sent in good articles, too. In January, 1995 the world got to see the premiere issue of *AIR*.

Those of you who have been following us on our *Irreproducible*—nay, *Improbable*—journey know that we have had other adventures. But we shall speak of them, if ever, anon.

What's What

Most of the articles in this book first saw the light of day or the dark of night in *AIR* or in *mini-AIR*. A few of them appeared before *AIR* was born. Most of the items are dated, in one or another sense of that word.

From Irreproducible to Improbable

The early history of the Journal of Irreproducible Results

by Harry J. Lipkin

Department of Particle Physics
Weizmann Institute of Science
Rehovot, Israel

During the early 1950s both Alexander (Leshek) Kohn and I, Harry J. (Zvi) Lipkin enjoyed writing humorous scientific articles. On April 1, 1955 Leshek composed an issue of a fake journal, the *Journal of Irreproducible Results,* Volume II, with Anonymous Editors and a cover containing a list of papers previously published in Volume I, which did not exist. Leshek was so pleased with the response that he decided to go public and start a real journal of scientific humor. He was a biologist and wanted a physicist collaborator to join him in starting this journal. A mutual friend introduced us.

Our first public issue, Volume III, July 1956, contained articles written by the two of us with various pseudonyms. Our editorial stated: "In view of the fact that the publication of the second volume . . . has drawn considerable interest in scholar and non-polar circles in this country and abroad, the Board of Editors has decided to cast its anonymity aside and to come out in the open with the publication of Volume III. It has also been decided to open wide the gates of the journal to all possible contributors of this and other worlds."

The next two issues, irreproducibly labeled Volume IV, January 1957, and Volume III, Number 2, April 1957, began to contain contributions from others in addition to the two of us. From this point it took off.

The first issues were typed by a secretary at Leshek's biological institute on a mimeograph stencil and reproduced by the institute's mimeograph machine. The next step in improved publication was the use of the multilith at the Weizmann Institute.

The first formally published and copyrighted *Irreproducible Results* appeared in the *Proceedings of the 1957 Rehovot Conference on Nuclear Structure,* which had a session entitled "Irreproducible Results"[1] containing some of my humorous articles and a commercial for *JIR.* This principal International Nuclear Physics conference in 1957 contained first reports of the exciting developments following the discovery of parity nonconservation. The *Proceedings* were therefore very much in demand and spread the word about *Irreproducible Results* and *JIR* to Nuclear Physics Institutes and libraries all over the world, where the book with its jokes are still available today.

However, one paper, "Religion, Thermodynamics and Communism," met a peculiar reception in the Communist world because it was making fun of communism, which was even worse than being anti-communist.[2,3] All *Irreproducible Results* were removed from the copy in the library of the Rumanian Academy of Sciences in Bucharest. Several years later the full unexpurgated version of the book appeared in the library of the Polytechnical Institute in Bucharest. It had been photocopied in its entirety by the Chinese, who evidently decided that the book was OK, because it made fun only of the Russians.

JIR flourished and accumulated more readers in the years 1956–58. In the academic year 1958–59 both Leshek and I were on sabbaticals in the U.S.

Harry J. Lipkin

and Volume VII of *JIR* appeared with a note that this issue comes from our temporary offices in the U.S.A. Volume VIII was back home in Israel. At that time Leshek and I invented the Society for Basic Irreproducible Research (SBIR) with Leshek as president and myself as secretary under the pseudonym X. Perry Mental. I still have my official membership card in this nonexistent irreproducible society. Volume IX began with the statement that *JIR* was edited by the Society for Basic Irreproducible Research.

Papers from *JIR* were translated into other languages and included in anthologies of scientific humor; e.g., *A Random Walk in Science* and the Russian book *Physicists Are Joking.*

The golden age of *JIR* came to an end after Volume XII, when the demand for subscriptions became so great that we could not cope with it. Leshek and I worked on *JIR* as a hobby; we did not want to

become businessmen. Leshek found a businessman who was ready to handle all the business administration and publication and leave us in full control of the editing. But this did not work. I saw the end coming and ceased to play an active role in *JIR* editing. Leshek fought to keep control of the editing in our hands and gave up.

In 1990, the *JIR* was sold after having lost almost all of its subscribers. The new publisher asked Marc Abrahams to become the editor of this magazine that he had never seen. Marc immediately contacted Leshek and the editorial board. Leshek became Marc's mentor and *JIR* limped along until 1994, when both the newer publisher and the original businessman made life impossible.

Shortly before Leshek's death, he and Marc attempted to preserve the original spirit of *JIR* by founding a new journal. Marc is continuing to edit and publish the *Annals of Improbable Research (AIR)*, which, as far as I am concerned, is the real successor to *JIR*.

Notes

1. Harry J. Lipkin, in *Proceedings of the 1957 Rehovot Conference on Nuclear Structure,* edited by H. J. Lipkin, North-Holland Publishing Company, Amsterdam, 1958, p. 391.
2. M. Octupolsky, E. Quadrupolsky and M. Dipolsky "The Expansion of the Communistic Field into Multipoland"; *Reviews of Modern Politics,* vol. 8, 1956, p. 45, and *Eastvestia Acad. Nauk,* vol. 4, 1955, p. 37.
3. "The Second Quantization and Renormalization of the Communistic Field, I. Removal of Divergences in the Leningrangian Formulation," *Dat. Man. Ys. Dedd.,* vol. 5, 1956, p. 38, reprinted in ref. 1, p. 608.

Kinetics of Inactivation of Glassware

by Alexander Kohn

This is the ground- and glass-breaking article that Alex Kohn wrote in 1955. It became the entire first issue of the *Journal of Irreproducible Results* (that first issue was of course labeled "Volume II, number 1," and this article makes reference to articles in the non-existent volume I) which Alex and Harry Lipkin edited together. Thirty-nine years later, Alex and I, joined by Harry and the rest of the now-sizeable gang, co-founded the *Annals of Improbable Research*. Alex passed away in late 1994, weeks before the first issue of *AIR* was mailed to subscribers. This article is reprinted with permission from Chana Kohn. The citations at the end are as they appeared in the original version.

Introduction

Since the times of the Phoenicians some facts about the peculiar properties of glass have been known; one of these properties is the high degree of breakability of glass products. Although much has been written about other properties of this substance,[1,2] about means of production[3,4] and methods of use,[6,7] its breakability has very seldom been mentioned. Such obvious means of smashing glass as were used by famous personages in history,[8] or throwing drinking glasses behind one's shoulder[9] are amply documented and will not be dealt with in this communication.

A survey of the situation concerning the availability and maintenance of glassware in various scientific laboratories showed that this subject needs reconsideration and more systematic research. This will be the subject of the present communication.

Materials and Methods

Glassware

We understand glassware to be a chemical product containing various proportions of CaO, $NaSiO_3$, Al_2O_3, ZnO, as well as oxidation products of other metals.[10,11] As the subject of this communication primarily deals with physical properties of glass, no chemical classification of this substance will be given. Of much greater importance in this research project was the form in which glass appeared. The following finished products were used: Petri dishes, test tubes, pipettes, Erlenmeyers, bottles and flasks of various volumes and forms, beakers, as well as specially constructed apparatus such as distilling flasks and condensers, connection tubes, Dewars thermometers, syringes, etc. One should also distinguish between products made of neutral glass, soda glass, and Pyrex. This definition is important as in those instances where the differences between these types of glass were not taken into account, the attempts to join by oxygen flame manipulation such two different types produced disastrous results.

In case of direct exposure of glass vessels to flame, great importance should be attached to knowledge of the type of glass used.

Half life

The rate of disappearance of glass may be defined by a term borrowed from nuclear chemistry: half life. This is the time required for a reduction to half of the functional number of a given species of glassware. Before this investigation was initiated it was generally assumed that the average half life of glass products was of the order of 5 to 10 weeks.

Methods of inactivation

These can be broadly defined as:

A. Mechanical. These methods may be subdivided as follows:

1. Shock inactivation
2. Vibration
3. Stress and pressure
4. Gravitational

The last method is the most frequently used and may be employed in two variants: normal gravitational field, and the centrifuge.

B. Thermal inactivation will appear under four headings:

1. Direct exposure to oxygen or hydrogen flame
2. Exposure to Bunsen flame, protected by asbestos grid or bare
3. Autoclave
4. Sterilizer—hot oven

C. Chemical inactivation is of rather minor importance. One should, however, mention the use of concentrated solutions of KOH or NaPH or of hydrofluoric acid.

D. Willful destruction.

Finally there exists one method not listed above, this being a combination of all three, A, B, and C, namely the thermochemical-explosive, which generally is unforeseen.

Results and Discussion

The study of the subject of inactivation of glassware indicated that of the various methods the most efficient and most frequently used are the thermal and mechanical, and in the latter group those due to gravitational effect. The normal gravitational inactivation is obtained when the glassware under consideration is simply intentionally or unintentionally suspended in air, preferably above a concrete floor or even a wooden table top, whereupon all support from below or above is withdrawn. It has been also reported that remarkable acoustic and "cursing" effects are obtained by tripping persons carrying trays with larger quantities of glassware. Splash effects with bottles or vessels filled with biological stains[12] or with chromic-sulfuric acid have also been recently obtained by the simple gravitational method.

The increased gravitational field obtained in centrifuges lends itself readily to inactivation of centrifuge tubes and bottles. There are, however, two requirements which should be strictly observed in order to obtain the desired effects. One of them is careful unbalancing of the two opposite cups. When this is done, a combination of vibration and gravity seals the fate of the tube—in most instances, of the

tube that contains the substance being purified by centrifugation. It has been reported from several laboratories that the responsible technicians in whose hands such inactivation occurred succeeded quite well in recovering the centrifuged substance by filtration of the slurry containing the glass fragments.

The other requirement is the introduction, into the cup, of a tube slightly longer than the space provided in the centrifuge for the tube. This requirement is essential only in the case of horizontal centrifuges and usually does not apply to conical centrifuges.

Stress and pressure methods will be dealt with separately under the headings of flasks, bottles, and syringes.

Rotem[13] has shown that evacuation of 1–2 liter Erlenmeyer flasks to approximately 1 mm Hg pressure resulted in a so-called implosion. This is entirely a pressure effect.

Thermal inactivation

There are several varieties of this method as listed above, and their effect on glassware will be described under separate headings. One method, however, which is extremely rarely used will be described here. The method was recently devised by Fendrich & Nir, and is described in detail here as no published account of it has appeared yet. Approximately 50 Pyrex tubes of 16×180 mm are carefully cleaned and dried; their mouths are wrapped in tinfoil to form a cap and the tubes are horizontally arranged in a large Pyrex rectangular dish. The whole assembly is introduced into an oven, thermoregulated to $300°$ C. The thermoregulator is slightly disarranged and the tubes left in the oven overnight. The next morning the oven is opened and a nicely fused mass of glass is found. The temperature is found to have risen to $800°$ C.

As to the inactivation by autoclavization, it has been found by several workers that satisfactory results are obtained with flat bottles normally used for culturing bacteria. Out of every 10 bottles introduced into the autoclave, 1–3 are taken out cracked or sometimes even split.

An original method of flask destruction based on both mechanical stress and thermal interaction was attempted in our laboratory by I. Hertman.[14]

He filled the bottles up to the rim with water at room temperature, closed them with a screw cap and introduced them for varying periods of time into the freezer compartment of a refrigerator. Only in rare cases was the bottle taken out without being cracked to pieces. This experiment neatly demonstrates the thermal expansion of water around 0° C.

Thermochemical explosive methods were introduced into this institute for the first time in 1955.[15] A suspension of West Nile antigen in acetone was introduced into the glass bowl of a Waring blender, and the instrument was started by depressing the electrical switch. The resulting spark exploded both the mixture and the bowl, and the flame spreading on the wall and table is reported to have reached even the eyebrows of one of the workers.

Not satisfied with such simple methods of starting thermochemical reactions, Sh. Miller[16] distilled acetone solution in vacuo over an electrical hot plate and required prolonged medical attention after the implosion of the distilling flask.

Pipettes

These may be inactivated mostly by the gravitational method. In order to induce gravitational breaking, the pipettes, in an aluminum tube, are set on the table top at an angle higher than 20° (Figure 1), the opening of the cylinder being directed towards the table edge. This results in an accelerated movement of the pipettes, which find their lowest energy position on the floor.[17] The number of pipettes later found on the floor is generally considerably higher than that originally in the cylinder.

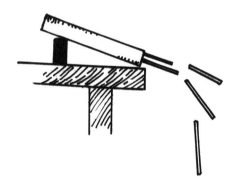

Figure 1: Gravitational action on pipettes.

Introduction of the pipettes into disinfection cylinders rarely results in inactivation, but the mass method of Zelda et al.[18] for a quick transfer of pipettes from cylinder to sink gave astonishing results.

Following the reports of Kellner[19] on photoreactivation of ultraviolated microorganisms by visible light, attempts have been made to apply a similar principle to inactivated pipettes. This is done only in cases where the damage to the pipette is terminal and does not affect more than 10% of the total length. The damaged end is first cracked with a dia-

mond knife and cut off. The reactivation is thermal, in an oxygen flame. When the damaged end is on the non-graduated side, the reconstituted product is as good as the new one. Efforts were made to reconstitute pipettes damaged at the graduated ends. This entailed glass-drawing, and the final products could be considered as a mutant of the Kimbell type. The accuracy of the reconstituted pipette was not always satisfactory.

Petri dishes

The rate of inactivation of Petri dishes is probably the highest of any type of glassware. The reason for this high rate of inactivation is to be looked for in the velocity of purification of Petri dishes of all residual materials, such as different types of agar media. The high turnover rate of Petri dishes in laboratories puts a great demand on the working facilities of the cleaning staff. Under this pressure, the velocity of washing increases to the point where the fingers of the handling personnel become slippery and a rich harvest of broken pieces of Petri dishes may be found in the waste bins.

Another method for the annihilation of Petri dishes is employed in cases where the lid of the copper casing in which the dishes are normally stored becomes stuck. The rather violent pulling off of the lid eventually results in the box bursting open and the dishes flying out and finding their lowest energy level on the floor (Figure 2).

Figure 2: Handling of Petri dishes.

Erlenmeyer flasks and beakers

This type of glassware is being continuously inactivated by the thermal processes. One of the most common methods is heating a flask containing agar medium over an open flame, and leaving the room for 2 minutes more than required for melting the medium (or asking somebody else to watch the flask). At a certain point in the melting process the excess of gases form a foam which, rising violently, bursts out, propelling the stopper in front of it. At such critical moments the person in charge usually grasps the neck of the flask with bare fingers, so that before the flask has the opportunity of safely reaching the table top, it is left to its own resources in mid-air. Here the forces of gravity take over and the result is as described above. In certain cases plunging the hot flask into cold water does the trick and leaves the neck in the hands of the experimenter, while the rest is eventually cooled. Although the method of holding beakers is a little different from that of holding flasks, the final results of handling them are also as described above.

Recently an Erlenmeyer flask was found tightly glued to a marble table top in the cold room. All attempts, chemical as well as thermal, to separate the two were in vain, and finally the flask had to be sacrificed and removed piece by piece.

A strange additive effect resulting in inactivation of glass should also be mentioned: When an attempt is made to remove sediment of organic matter from the bottom of a glass vessel by means of a glass rod, the general result is a hole in the bottom of the vessel, i.e., glass + glass = lack of glass = hole.

Syringes

Syringes, as is well known, consist of a barrel and a piston. In the Luer type used in this institute, both parts of are made of glass, so at least one does not encounter the type of inactivation frequent with Record syringes where hot-oven sterilization results in ungluing the metal and glass parts. It is very unfortunate that the fit of the barrel and piston are so much individualized that once one of them breaks the other cannot be used any more.

The breakage of syringes occurs as a result of two manipulations:

a. Leaving the syringe without cleaning it properly. This eventually results in a 100% tight fit, such that the plunger can no longer be removed from the barrel.

b. Testing the tightness of the syringe, when done by occluding the tip with a finger, then pulling up the plunger so as to create a vacuum, then releasing the plunger, results in the neat removal of the front part of the barrel. The barrel may then be used as a connection tube for rubber tubing.

When syringes are sterilized in a boiling water bath, it is advisable to leave the bath unattended for several hours. When the bath is electrically heated, after evaporation of the water the rising heat will melt the solder joints. At such a point one may expect the return of the attending person or arrival of somebody else attracted by the smell. The situation is normally handled by cooling the whole business, i.e., pouring cold water into the container and over the syringes. The cracking patterns on the barrels are most interesting to behold.

Special instruments

A very original method of inactivating the Warburg manometer has been described by Avi-Dor (personal communication). The Warburg apparatus produced in the U.S.A. is built to be operated by individuals of 5′ 7″ or more. Some of our Israeli scientists, being of lesser stature, have to insert the cup connected to the manometer into the water bath while looking at it from below (instead of above). As a result of a paralactic error, the cup is often gently banged against the rim of the bath and the manometric tube broken. The breakage generally occurs at the same spot on the capillary. Dr. Avi-Dor believes that the apparatus is designed by the producer so that when something breaks it will not be the cheapest part.

Glass beads

This glass product, being composed of discrete entities, lends itself very well to inactivation. The methods of inactivation are, however, quite different from all those described above. The rate of disappearance of glass beads from scientific laboratories very much resembles the rate of evaporation of solvents such as ether or acetone, with the difference that temperature is not a factor in the velocity of disappearance of glass beads.

An interesting observation was made by Kohn and Zelda.[20] Assuming that the law of preservation of mass also applies to glass beads, they searched for the missing fraction in various parts of the Institute. Individual beads were found under work benches and in drawers, but the only place where a large number of beads was discovered was the bottom of a W. C. syphon opposite the wash room.

Glass windows

Although this product does not belong to the category of laboratory equipment, it will be shortly considered here as that part of the laboratory that permits the light to enter and prevents rain from diluting reagents kept on window sills.

The greatest contribution to inactivation of window panes has been made by Greenberg and his collaborators.[21] They specialized in the frosted glass type. It should also perhaps be mentioned that they also have done the most work on the problem of reactivation of window panes.

Interesting experiments were made by R. Ben-Gurion.[22] While attempting to destroy rich fungal flora on a marble window sill covered with agar residues, she heated the growth with a Bunsen flame. The uneven heating of the window pane eventually forces the maintenance department to replace it with a whole one. The same author also demonstrated that it is possible to inactivate a bus window pane by appropriate elbow pressure.

Kinetics

The calculations of the rate of inactivation of glassware are based on the formula:

$$Q_T = Q_0 \cdot e^{-KT}$$

where Q_0 is the initial quantity of glassware, Q_T the quantity after period T, and K = the exponential constant, which was found to vary with each species of glassware. For Petri dishes it was found to be $K = 0.06$, when T is measured in weeks (Figure 3).

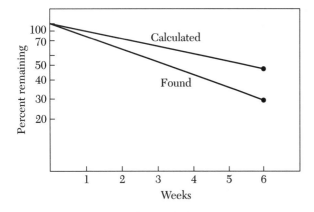

Figure 3: Inactivation of Petri dishes.

One factor, which we call the uncertainty or mystery factor, introduces an error, however, so that the calculated rate of inactivation does not correspond to the experimental one (Figure 3). This discrepancy is especially true for Petri dishes and culture bottles. No experimental method for determination of the mystery parameter has been devised yet.

It has been pointed out the inactivation of glassware is not a physical process but a chemical one.

Similarly to breakage of diamonds, breakage of glass is probably due to destruction of chemical bonds in large molecules. One should, therefore, employ the kinetic formula of Arrhenius

$$V = A \cdot e^{-E/RF}$$

where E is the energy of activation and A the collision factor (it is obvious from the data presented in this article that in this institute E is low and A very high).

Summary

The data collected and presented on inactivation of glassware show that there exist more factors affecting the inactivation than hitherto suspected. The various parameters are calculated and discussed.

Acknowledgments

The author is indebted to Y. Zelda and her staff, without whose faithful collaboration the volume of this article would have dwindled to that of a short note.

The author wishes to thank D. Yasky for the calculations, the Chief Administrator of the Institute for permission to publish this article, and himself for the bright idea of writing the article.

Notes

1. Silverman, A. Glass evolution: A factor in science. *J. Chem. Educ.*, 1955, 32, 149.
2. Richter, E. M. A. The room of ancient glass. *Bull. Metropolitan Museum Arts*, 1911.
3. Morey, G. W. *Properties of Glass*, 2nd ed. Harcourt, Brace & Co., N.Y., 1954, 336.
4. Pliny. *Historia Naturalis*. Johannes de Spira, Venice, 1469, 36, 26.
6. Mumford, L. *Technics and Civilization*. Harcourt, Brace & Co., 1935, 126.
7. Hoverstadt, H. *Jena Glass and Its Scientific and Industrial Application*. MacMillan Co., N.Y., 1902.
8. Socrates and Xantippe, *Review of Antiquity*, 1888, 6, 5.
9. *Secret Reports on the Banquet Held in Royal Palace*. Moscow, 1772, 15, 1.
10. Eitel, W., Pirani, M., and Scheel, K. *Glasstechnische Tabellen*. Springer Verlag, Berlin, 1932.
11. Neri, A. *L'arte vetraria*. Stameria de giunta, 1612.
12. Segal, Z. Efficient dispersal of methylene blue from flowmeters. *J. Irreprod. Res.*, 1955, 1, 25.
13. Rotem, Z. Preparation of autonomous vacuum system. *J. Irreprod. Res.*, 1955, 1, 45.

14. Hertman, I. Supply of cold drinking water in summer. *Bull. Isr. Inst. Biol. Res.,* 1954, August.

15. Not to be published.

16. Miller, Sh. Distillation of acetone in vacuo. *J. Irreprod. Res.,* 1955, 1, 2.

17. Kohn, A. Improved method of storing sterile pipettes for use. *J. Irreprod. Res.,* 1955, 1, 67.

18. Zelda, Y. et al. Methods in washing and sterilizing glassware: II. Mass-accelerated transfer of pipettes from chromic-sulfuric acid to water. Unpublished results.

19. Kellner, M. *J. Bact.,* 1949, 58, 511.

20. Kohn, A. and Zelda, Y. Report on the search for glass beads. *Bull. Isr. Inst. Biol. Res.,* 1955, 1026.

21. Greenberg, Y., Alkuser, Ch., Goldenberg, Sh. and Wolf, I. Annual report of maintenance crews. *Bull. Isr. Inst. Biol. Res.,* 1953, 1001.

22. Ben-Gurion, R. Fight against fungi. *Bull. Isr. Inst. Biol. Res.,* 1953, 1003.

CHAPTER 2

Nobel Thoughts

I have been interviewing Nobel Laureates for six years now, wasting, all told, an impressive amount of time and thought from people who have better things to do. Then again, a surprising and heartwarming amount of wisdom has slipped into the responses.

The Laureates often turn their great minds, unafraid, to the problems of everyday life. How to deal with junk mail, whether to use a pencil or a pen, how to distinguish between fatheads and phonies, whether one should read on the toilet—these are all questions that the average Joe has tried and failed to answer.

While most people are embarrassed to admit "wasting" time on such things, Nobel Laureates view these problems as worthy intellectual challenges.

More surprising, maybe, than the fact that these scientists are indeed human, is that they have a sense of humor. Some of them are lifelong pranksters, punsters, performers and storytellers. Further evidence, for those who need it, is in Chapter 3, which describes the various Ig Nobel Prize ceremonies.

And most surprising of all, to anyone who thinks that great scientists are cold, superhuman reasoning machines, is another fact. Throughout their careers, and especially since receiving the highest official honors the world knows how to give, almost all of them have worked their tails off trying to get people of all ages and abilities interested and involved in science. Some of our favorite mail at *AIR* comes from parents and teachers who have seen their children get excited by these strange little discussions.

When each new interview is published in *AIR*, the headline reads "Nobel Thoughts: Profound Insights of the Laureates." I conducted about half of the interviews here in person, the rest by telephone.

NOBEL THOUGHTS:
James Watson

This appeared in 1993.

James Watson is president of the Cold Spring Harbor Laboratory. In 1962 he and Francis Crick shared the Nobel Prize in Physiology or Medicine for discovering the chemical structure of DNA—the now famous double helix. We spoke while striding briskly down the middle of a street.

What is your secret for staying awake during a dull lecture?

Read a newspaper.

What is your technique for dealing with overeager admirers?

Shyness.

How do you respond to people who ask you foolish questions?

Politely.

Do you have any advice for young people who are entering the field?

Enter an underpopulated field. Work with young people, not established stars.

NOBEL THOUGHTS:
Roald Hoffmann

This appeared in *AIR* 1:6 (November/December 1995).

Roald Hoffmann is the John A. Newman Professor of Physical Science at Cornell University. In 1981 he and Kenichi Fukai shared the Nobel Prize in Chemistry for their theories concerning the course of chemical reactions.

How do you deal with junk mail?

Junk mail is the mail that gives me the greatest pleasure in the world, because I know immediately what to do with it.

You never read a single piece of it?

Oh, I read some of it.

Which do you read and which don't you?

Well, if they talk about sex after fifty I look at it. I'm also hoping always that someone is going to give me money for my research. Sometimes that gets hidden in the junk mail envelopes, you know.

Then you don't make any effort to prevent junk mail from coming in?

No, no. I love it. I told you, it's the greatest psychological pleasure to throw it away.

What are your secrets for discouraging unnecessary paperwork?

It depends. I have different strategies. When people want me to join editorial boards of journals, I suggest that they offer a free subscription to 200 libraries in underdeveloped countries. That usually gets them off. When somebody writes to me and asks me to write a book for them for some commercial publication, I usually offer them a deal where I'll do that in exchange for a copy of every book they've published that year. I actually like suggesting these unreasonable deals. It usually gets the commercial people

off your back. It's very hard to get the American Jewish Congress or the American Cancer Society off. I haven't quite figured out how to do that.

What's your technique for handling salesmen who call you on the telephone?

Well, I often tell them I have a family policy that we don't deal with solicitation by telephone, and that if they call again I'll keep track of them in a book and not give them any money. There are junk bond people who call, people who want me to invest money. I tell them to write to me and that if they write to me by hand I'm more likely to read the letter.

And then do you read the letter?

Yes.

And do you do anything beyond reading it?

I throw it away.

What advice do you have for young people who are entering the field?

I think they should volunteer to teach freshman chemistry, because it is through that teaching that they'll become better researchers. I got A+ grades in my thermodynamics courses in college and grad school, but I didn't understand thermodynamics until I had to teach it to first year students. What you learn from first year students is strategies for simplifying. Also, you learn strategies for explaining.

I believe that research and teaching are a continuum in which there is just a matter of dealing with different audiences. And that audience of first year chemistry, a thousand kids—and I've been teaching it all the time since I've been here—you're explaining it to a mixed audience some of whom understand it, some of whom don't. That's exactly what you're doing when you're writing a paper. People pick up the journal and some of those people understand what you're saying and some don't. If you learn the strategy for communicating to that audience of a thousand, you will learn the strategies of communicating in science in general.

I feel that people in industry are at a disadvantage [in this way]. Their equivalent is the presentation to their manager, which is also an audience roughly equivalent to that of a freshman chemistry class. But if they don't have that, some of the people who are shielded in industry in the research-oriented jobs don't learn how to explain their stuff. Now, for the one-half of one percent of us who get by on our brains, that's okay, but for the rest of us we've got to somehow sell our things. I really think teaching is a good way to do that.

NOBEL THOUGHTS:
Dudley Herschbach

This appeared in 1991.

Dudley Herschbach is the Baird Professor of Science at Harvard University. He, Yuan Lee, and John Polanyi shared the Nobel Prize in Chemistry in 1986 for their work on atomic tunneling phenomena.

Which do you prefer, pencils or pens?

Pens, although I use both indiscriminately. I use an old fashioned pen I actually fill with ink, so I often have ink stains. It's the only thing I can write neatly with so other people can read it.

Is it important to write on lined paper?

No. When I'm doing math, of course, I prefer unlined paper. I probably use equal quantities of lined and unlined paper.

What characteristics are most important when you are buying a notebook?

I like to have what seems to me an appropriate size. For most of my work, this (11″ × 8 ½″ by ½″) is the right size. On the other hand, the notebook I use for phone numbers and messages has got to be this size (9″ × 6″ × ½″). I can't tell you why. One of these small notebooks will usually run me for, oh, two years. [Professor Herschbach's notebooks are all spiral bound and contain lined paper.]

Do you have any advice for young people who are entering the field?

There are two ideas I think it's important to get across to them.

The first is that to be a scientist is not a matter of special talent. So many kids lack self-confidence; they're always too impressed by other people. The image of Einstein, the solitary dominating genius, is an unfortunate one in the minds of young people. Potentially modest gifts are quite adequate, provided that you love it. It's a matter of how much you want to get there. In so many other areas of human affairs it's not that way at all—the timing can be important if you're in real estate, for example. But science is different. I prefer to think of it as a lovely damsel on a mountaintop, waiting for you to find the way to the top of the mountain. This damsel is patiently waiting for you. She's giving you encouragement from time to time by dropping little plums in your lap. Ordinary human talent over time can get you there.

One thing that frightens students is the feeling that you've got to get it right. But science, in contrast to most activities, lets you get it wrong a lot of the time. Much of the time you're confused. The scientist is elated by being confused. You know you've got to have this period of confusion, and you trust that if you keep after it, what you're after will take shape. And what turns out to be really exciting is not what you had thought would be exciting.

Being a scientist is like being a musician. You do need some talent, but you have a great advantage over a musician. You can get 99% of the notes wrong, then get one right and be wildly applauded.

The second thing is that science for the most part is cooperation among a great body of people. There really isn't that much competition. That's very different from many other human activities. Many times, it's almost embarrassing that you don't make much of a contribution yourself, except to recognize several things that other people have done and put them together in ways that the other people hadn't even thought about. You can be pretty damn inept in many things and yet be pretty good in science, because you benefit by what other people have done. Many times, people who are really artistic in temperament can be pretty good scientists. Science is a very congenial, friendly thing. The sad thing is, it doesn't look that way to other people.

NOBEL THOUGHTS:
Richard Roberts

This appeared in *AIR* 1:5 (September/October 1995).

Richard Roberts is the director of research at New England Biolabs. In 1993 he and Philip Sharp shared the Nobel Prize in Physiology or Medicine for the discovery of split genes.

How have fatheads and phonies affected your education?

There was one classic fathead who attempted to teach me physics, but in fact he ended up having a personal vendetta. When I decided that playing billiards was more interesting than physics, he took offense, and spent the rest of the time trying to catch me playing billiards. He eventually did. I loathed this man with a vehemence. We used to call him "slug," because he had greasy black hair. He always wore black. One was convinced there was a trail of grease behind him. A particularly obnoxious fathead.

What became of him in later life?

Well, he stayed on at school for a while and then they hired a new headmaster, and the new headmaster took an instant dislike to him. He went off and did other things. Unfortunately, neither the new headmaster nor myself ever tried to find out what happened to him, since we didn't want to know.

How about fatheads who had an effect on your career?

Oh, God. . . . No, I can't say that. . . . If I told you the story it would be obvious who it was, whether I named the name or not, and he would be offended, so I don't want to do that.

I can tell you a story about a phony. I started life as a chemist, so when I decided I wanted to do molecular biology, there were a lot of people in the

Roberts (left) takes a star turn in "The Interpretive Dance of the Electrons." The ballet had its world premiere (and only performance) at the 1994 Ig Nobel Prize Ceremony. Visible at right is Nobel Laureate William Lipscomb.

chemistry department—many of whom actually I guess one could call phonies—who thought that there was no such thing as molecular biology. They said there's chemistry and there's biology, but there's no such thing as molecular biology. And so, of course, that only made me want to be a molecular biologist more and more and more. And so they had a very positive influence on me, because as soon as I heard them saying that, that's immediately what I wanted to be.

In your experience in the fields in which you've been working, are there more fatheads or more phonies?

Well, there are loads of phonies. There are really plenty of phonies. But I think marginally the fatheads have it. I would say by a fat head they have it, probably.

A two-part question: What's your prediction about the effect that fatheads and phonies will have on your future career?

That's a very tricky one. The fatheads almost certainly will have an effect on my future career, but I can't imagine quite what. And phonies—they're everywhere. One cannot fail but be influenced by phonies. The trick, of course, is to spot them as phonies early on, and then ignore them. As soon as you spot a phony, ignore him immediately.

What do you hope to teach your children about how to spot fatheads and how to spot phonies?

My children are eight and five. They're well grown. They've already figured out how to spot fatheads and phonies. In fact, they're much better at spotting phonies than I am.

What advice do you have to young people who are entering the field?

Do it. Do it quickly, while the field is still as exciting as it is now.

NOBEL THOUGHTS:
Mel Schwartz

This appeared in 1992.

Mel Schwartz is Professor of Physics at Columbia University, where he teaches freshman physics. In 1988 he, Leon Lederman, and Jack Steinberger shared the Nobel Prize for Physics for their work in discovering the muon neutrino.

Do you buy new cars or used cars?

I only buy new cars, because I don't want other people's problems. Besides, when I get rid of my old cars, I never tell anybody about the problems.

Do you like to drive?

Unfortunately I have no choice. I spent 25 years in which I drove 100 miles a day. That was 500 miles a week . . . 25,000 miles a year . . . 625,000 miles in the last 25 years.

Did I like it? I don't remember a damn thing about it. I don't know if I like it. I just have to do it. I've actually spent 125 working days sitting behind the wheel of a car getting to work.

But I'll tell you what I like. I like driving in New York City, because I can act like a New York City taxi driver—but only when I'm using my wife's old car.

Do you prefer driving with the windows open or closed?

Closed. I don't want somebody reaching in and grabbing me by the tie in New York City. Windows closed, and doors locked.

Do you have any advice for young people who are entering the field?

Be aggressive, because in the end that's the only way you get ahead. That's meant partly in jest, but it's partly true. Try to make an impact so at the end people will know that you lived.

NOBEL THOUGHTS:
David Baltimore

This appeared in 1992.

David Baltimore is president of the California Institute of Technology. In 1975 he, Renato Dulbeco, and Howard Temin shared the Nobel Prize for Physiology or Medicine for their discoveries concerning the interaction between tumor viruses and the genetic material of the cell.

When you were a boy, did you read comic books?

Yes. *Superman*, *Batman*, and the like.

When you read a daily newspaper, which sections do you read first?

The front page. And the last thing I would ever turn to would be the sports section. In between, I graze.

Do you recommend that people read in the bathroom?

Absolutely. It's one of the few times that you can really focus your attention—on what you're reading. In fact, one of the problems with the design of bathrooms is that there's generally not a good place you can put your book when you get up from the toilet.

Do you have any advice for young people who are entering the field?

Even if it seems hard and unremunerative, stick with it—it's great fun.

NOBEL THOUGHTS:
Linus Pauling

This appeared in 1993.

Linus Pauling is widely regarded as one of the giants in the history of science. He has been called the father of modern chemistry, and his pioneering inquiries have ranged wide and far in the disciplines of biology, physics and medicine. Pauling is the only person who has received two undivided Nobel Prizes. In 1954 he received the Nobel Prize in Chemistry for his work on the nature of the chemical bond and its application to the structure of complex substances. In 1962 he received the Nobel Peace Prize for his efforts to bring about the treaty banning tests of atomic explosives in the atmosphere. Pauling died in 1994.

To what extent did your schooling interfere with your education?

I don't think it interfered at all. I think I was fortunate going to public schools in eastern Oregon and then in Portland. They were excellent schools, grammar school and high school.

To what extent did you interfere with your education?

Very little. Only one episode that I remember. After I'd been in high school for three years and a half, having started in February—mid-year, you see—I realized that I could go on to Oregon Agricultural College if I graduated at the end of the term. There was a requirement that to graduate high school the student needed to have two terms of American history. I was always interested in history, so I signed up for American History A and American History B. The teacher who was registering said I had to get the permission of the principal. I went to the principal, and he said, "No," so I turned around and went out and changed the two terms of American history to seventh semester mathematics and eighth semester mathematics—trigonometry was one of them, and advanced algebra—changed my schedule and didn't get a high school diploma. So I interfered with the system to that extent. Then twenty-five years later, perhaps, I was given an honorary high school diploma by petition of the high school students.

What is the most intriguing experiment someone might do regarding human nature?

I don't think I could answer such a question without thinking awhile. I have tried to.

Do you have any advice for young people who are entering the field?

Well, I have advice for young people in general. That's a question I get asked reasonably often. I say you should look around carefully at the members of the opposite sex, and pick one out that you'd like to be with all your life. Get married young, and stay married. Then second, I say try to decide what you like to do best—what you enjoy doing—and then check up and see if it's possible for you to earn a living doing it.

Is there a third point?

No.

Each year we present Ig Nobel prizes to people whose achievements cannot or should not be reproduced. Who would you nominate to win an Ig Nobel prize?

Well of course I'd be pleased to have [Edward] Teller get a second Ig Nobel prize so he could become listed in the Guinness Book of Records as the person who's achieved the most Ig Nobel prizes. [Editor's note: Edward Teller, the father of the hydrogen bomb and the foremost proponent of the "star wars" missile defense system, was awarded the 1991 Ig Nobel peace prize. The citation said that Teller had "changed the meaning of peace as we know it."]

Anyone else come to mind?

Well, let me see. In personal science, Dr. Victor Herbert I think deserves such a prize. He was at Hahnemann University and got fired because he got in a fistfight with the dean. He—Victor Herbert—is considered to be a great authority on vitamins, always testifying on vitamin cases, and he was on the food and drug board that National Academy president Frank Press fired when they brought in their report that the RDA's [Recommended Daily Amounts] be decreased. Then when the National Academy of Sciences had a new committee and got out a new report, he sued them for using some material that he had written—for plagiarism. I think that case has been thrown out of court.

And he in a sense is responsible for my having spent more than 20 years in this vitamin field. He irritated me so much about 1969 that I sat down and wrote my book *Vitamin C and the Common Cold*. Well, Victor Herbert is famous among orthomolecular nutritionists and physicians. You expect the Food and Drug Administration to be quoting him by just reading the reports, so they quote him as authority for statements that I think are just not true. Mr. Herbert seems to me to be a really good candidate.

Anybody else?

Well, there's an anonymous referee for *Physical Review Letters* who said that a paper that I wrote should be turned down, a paper in which I talked about the cluster of nucleons revolving about a central sphere. He said a structure of that sort is impossible because quantum mechanics requires that the normal state (or any other state) be either symmetric or antisymmetric. So I wrote to the editor and said: "Here, this fellow doesn't understand quantum mechanics, and you're using him as a referee! He would say that a molecule of hydrogen chloride, for example, couldn't exist." I didn't get any reply to that from the editor.

NOBEL THOUGHTS:
William Lipscomb

This appeared in *AIR* 2:2 (March/April 1996).

William Lipscomb is the Abbot and James Lowell Professor of Chemistry Emeritus at Harvard University. In 1976 he received the Nobel Prize in Chemistry for his research on the structure and bonding of boron compounds and for his general discoveries about the nature of chemical bonding.

In your paper "Boron arrangement in B_9 Hydride," you discuss two plausible hydrogen atom arrangements.

Well, the fact is, this was a very difficult problem, because it was presented to us as a compound containing eight boron atoms, when it really contained nine. We were completely at a loss to understand it, and we tried out all the possibilities for the B_8 hydride, and finally concluded that there couldn't be eight—there must be one more. So we put another one in and it worked. This is an illustration about science that if you eliminate all the other possibilities, then if only one more remains, that must be the correct one. Well, that reminded me—I was then at the University of Minnesota and a member of the Baker Street Irregulars, a chapter of the Sherlock Holmes Society "The Norwegian Explorers"—

Yes, Sherlock Holmes is very interesting. Now, we were talking about the hydrogen atoms . . .

Yes, I just wanted to illustrate that this was done using the method of Sherlock Holmes. In the complete works of Sherlock Holmes there are four places where Holmes says, "Wherever all other contingency fails, whatever remains, no matter how improbable, must be the truth." This is Holmes's method. Now, the other three places use wording that's rather like that but not exactly like that.

Indeed. Now about the arrangement of those hydrogen atoms . . .

The whole thing has to do with the number of boron atoms and the method of Holmes. This is a scientific method. You get all the possibilities except one final remaining one, and in doing this you have a method of science. And this is the method which Holmes used in at least four different places in the complete works of Sherlock Holmes—

About those hydrogen atoms . . .

You see, the Baker Street Irregulars is a society that has existed for a great many years, and they do research on Sherlock Holmes, finding problems to solve. Here's an example of a research problem. In the complete works of Sherlock Holmes, Watson's wife, who was born Mary Morstan, refers to her husband as John. Of course, that's his name—John H. Watson, as Holmes writes in the volume. And the

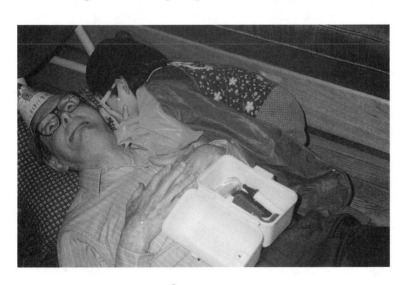

William Lipscomb with his daughter Jena.

solution of this problem was actually achieved by Dorothy Parker. There's one place in which he's referred to as "James," rather than "John." Now, it's impossible that she could have not known her husband's name. So the solution of this problem is that the H. stands for "Hamish," which is Scottish for "James." This is the kind of problem that one solves in Sherlock Holmes.

And the hydrogen question?

Yes. Well, Holmes is a model for scientific investigation. He worked in the days when they didn't use fingerprints. They used mental effort alone. They were presented with a problem, and he used his brains to work it out, and look for all the clues that he could. Now, what is a better way to do science?

Exactly. Now about those hydrogen atoms. They form an icosahedral fragment.

You see, the stories were not really written by Conan Doyle, who was a literary agent. They were written by Watson, except for four of them which were written by Holmes himself. So there's something wrong with the idea that these are to be attributed to Doyle.

Indeed. Now, given that each boron atom supplies four orbitals but only three electrons—

There's another aspect to Holmes that I really enjoyed very much. He was a violinist—a very fine musician, as Watson notes. Holmes needed some relaxation on a few occasions, and he used to go to concerts, and he played the violin. I really like that, because solving problems is an art, too, just like the performance of music, so it helps enormously.

Yes, it does. Well, thank you for talking with us!

You're very welcome. There's another method of Holmes that has not yet appeared in the literature, and that I hope to use. It refers to Silver Blaze. Silver Blaze is the name of a racehorse. One night the horse disappeared from the stable. Holmes was given the job to find out who did it, and towards the end of the investigation, he's discussing the case with Watson, and he says, "Well that's all right, but what about the curious incident of the dog in the night time?" And Watson says, "But the dog did

In addition to being a chemist and an avid reader, Lipscomb is an accomplished clarinetist. This action close-up of his hands was taken at the Fifth First Annual Ig Nobel Prize Ceremony, during which her performed a duet with jazz harpist Deborah Henson-Conant.

nothing in the night time." "That is the curious incident," Holmes says. That means, you see, that since *nothing* happened, that was an important clue that the dog recognized the person who had visited the stable. When we have an experiment and we expect something to happen—yet nothing happens—this is the appropriate reference from Sherlock Holmes. I hope to use that some day in a paper.

Reference

"Boron Arrangement in a B_9 Hydride," Richard E. Dickerson, Peter J. Wheatley, Peter A. Howell, William N. Lipscomb, and Riley Scaeffer, *The Journal of Chemical Physics*, vol. 25, no. 3, Sept. 1956, pp. 606–607. Footnote 4 of the paper quotes Sherlock Holmes describing his theory of contingencies. The quotation is taken from the Arthur Conan Doyle story, "The Bruce Partington Plans."

Nobel Thoughts:
Sidney Altman

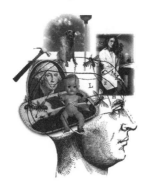

This appeared in *AIR* 1:1 (January/February 1995).

Sidney Altman is Sterling Professor of Biology at Yale University. In 1989 he and Thomas Cech shared the Nobel Prize in Chemistry for their work with ribonuclease-P, demonstrating that RNA not only can carry information but also can catalyze chemical reactions.

Could you discuss the relative merits of beer and potato chips?

In this world of concern for the environment, I think these two items have to be discussed with respect to the detritus produced. In that regard, I think beer is a hands-down winner. Not only can it provoke a variety of fluids to be emitted from its consumer, it can leave a lingering, potent smell in the area in which it has been used, whereas potato chips generally, as a vestige of their usage, leave behind a certain amount of cholesterol within the consumer, and occasionally, the uncomfortable feeling of crumbs or little pieces of crunchy material in one's bed or sofa or wherever the stuff has been consumed. So as a generator of jobs for those people who are interested in cleaning up the environment, and in general concern and awareness of this pollutant, beer is a hands down winner.

How many chips per glass of beer are optimal?

Well, that question is typical of a non-scientist. There is no regard whatsoever to quantitative observation or to matters of scale. What kinds of glass are we talking about? What sizes? What sizes of chip are we talking about? What density of chip?

Go back to the lab.

Do you recommend potato chips over pretzels?

This is also a question of such broadness that it begs an answer, because we have to remember that, these days, chips are not simply potato chips.

It's quite easy these days to get one's hand on chips made from all kinds of tubers, vegetables, fruits. And similarly with pretzels. The word pretzel simply has a connotation of something that is knotted, although if you look carefully at most pretzels sold anywhere these days, they are in fact not knotted. So it depends on what one's interest in topology is, and what one's aesthetic sensibilities are—for example, sweet potato chips are quite beautiful in terms of the colors they have.

So once again, this is a matter of individual choice and has nothing to do with science.

Do you have any advice for young people who are entering the field?

Watch out for the cow pats.

Every year we award Ig Nobel Prizes to people whose achievements cannot or should not be reproduced. Whom would you nominate for an Ig Nobel Prize?

The person who is conducting this interview.

CHAPTER 3

Ig, Ig, Ig Nobel— A Different Kind of Prize

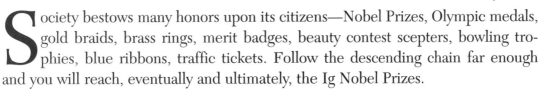

Society bestows many honors upon its citizens—Nobel Prizes, Olympic medals, gold braids, brass rings, merit badges, beauty contest scepters, bowling trophies, blue ribbons, traffic tickets. Follow the descending chain far enough and you will reach, eventually and ultimately, the Ig Nobel Prizes.

Ig Nobel Prizes are awarded to individuals whose achievements "cannot or should not be reproduced." The criterion covers a lot of ground. Beginning in 1991, we have been presenting ten prizes each year in the fields ranging from physics to chemistry to peace to art. Through happenstance, each crop of winners includes people whose achievements are (at least in retrospect) whimsical and wonderful, and other people whose achievements are perhaps not so wonderful. Many of the winners either come to the ceremony or send acceptance speeches via videotape or audiotape. One winner even went so far as to pay his own way from Oslo, Norway to Cambridge, Massachusetts to accept the prize.

If you win an Ig Nobel Prize, there is good reason to attend the ceremony. The prizes are physically handed out by genuine Nobel (not Ig Nobel) Laureates, in a lavish, oxymoronic ceremony at Harvard University's grand old Sanders Theater. Enthusiastic, oddly-garbed, standing room-only crowds applaud the new winners and deluge them with paper airplanes, and cheer on the Nobel Laureates as they good-naturedly take starring roles in bits of opera, ballet, or other improbable works of what some might be tempted to call art. The event is televised live over the Internet, and recorded for later broadcast on National Public Radio's *Talk of the Nation: Science Friday* and the C-SPAN television network, and is covered by newspapers, radio, and TV news organizations from around the word, as well as by all of the major science journals.

We held the first ceremony in 1991 at MIT, in a space that held 350 people. The night of the event, we didn't know if anyone would turn up. We were pleasantly surprised. Four Nobel Laureates showed up wearing Groucho glasses and fezzes, and people were crawling the walls of the building in an effort to be part of

Nobel Laureates (left to right) Sheldon Glashow, Eric Chivian, Dudley Herschbach and Henry Kendall handed out the Prizes at the First Annual Ig Nobel Prize Ceremony, which was held in 1991 in an old room at MIT. Photo: Roland Sharrillo.

the audience. The ceremony has grown every year. In 1995, it moved up the street to Harvard, where the Seventh First Annual Ig Nobel Prize ceremony will occur in October.

The ceremony has many parts, ranging from the traditional Ig Nobel Welcome, Welcome speech to the traditional Ig Nobel Goodbye, Goodbye speech. It has been described as being a combination of the Academy Awards ceremony, the Nobel Prize ceremony, a three-ring circus, and the old Broadway show *Hellzapoppin*.

The Welcome, Welcome Speech

The traditional Ig Nobel Welcome, Welcome Speech was delivered by Lois Malone. Here is a complete transcript of her speech:

Welcome, Welcome.

The Goodbye, Goodbye Speech

The traditional Ig Nobel Goodbye, Goodbye Speech was delivered by Lois Malone. Here is a complete transcript of her speech:

Goodbye, Goodbye.

The evening begins with a grand parade of the official audience delegations, representing groups such as the Junior Scientists (2nd graders from the Boston Public Schools); the Harvard Computer Society; the Museum of Bad Art; Lawyers for and Against Biodiversity; the Friends of Daryl, Daryl, Daryl and Daryl; and the Non-Extremists for Moderate Change (from Finland). The King and Queen of Swedish Meatballs make an appearance and, in the best tradition of roayalty, spend the evening doing and saying absolutely nothing. The gold-body-painted human spotlights appear to illuminate the proceedings. The majordomo and minordomo rush about officiously. Important people deliver the 30-second Heisenberg Certainty Lectures. Some lucky ticketholder wins the annual Win-a-Date-With-a-Nobel-Laureate Contest. Plaster casts of the left feet of Nobel Laureates are auctioned off. The audience heckles and throws paper airplanes. The people on stage return the heckles and throw back the paper airplanes. And, oh yes, the ten new Ig Nobel Prizewinners are announced, and their acceptance speeches delivered.

There is a serious side to this, more or less. The Ig, as it is known to one and all, is a celebration of science, and a demonstration that scientists really do enjoy their work and that science really is a warm,

Audience delegations enter the hall at the start of the Fifth First Annual Ceremony in 1995 at Harvard University's Lowell Lecture Hall. The cat at upper right is part of the evening-long semi-random slide show. Photo: Stephen Powell.

A woman in a cow suit cashes in her lucky ticket at the 1996 Win-a-Date-With-a-Nobel-Laureate Contest. Dudley Herschbach, that year's prize, insists that she sit with him on stage. Photo: Stephen Powell.

human, and wonderfully quirky endeavor—and that it's not such a terrible thing to try out wacky ideas. Most of what are now called the great discoveries in science were hooted and scoffed at when they were new. At one time, doctors who said you should wash your hands before doing surgery were thought to be cranks. Yet, nowadays most hospitals have surgeons who wash their hands.

Upon first encountering the Ig, some people wrongly assume that the point is to spotlight bad science. That is not at all the main point, and in most cases it is not the point at all. Many of the Ig Nobel Prizewinning achievements are really good science, and perhaps even of great benefit to humankind. Others are perhaps not—but hey, we all make mistakes. The Ig Nobel Board of Governors is always conscious of the harm done by the Golden Fleece Awards that US Senator William Proxmire used to hand out. Proxmire's gang was on the lookout for people to ridicule and pillory, and they sometimes

went after people whose work, though odd-sounding, was deserving of praise and support. The Igs, on the other hand, illuminate the endearing—and yes, perhaps odd—aspects of raw human striving. As mentioned above, many Ig winners are thrilled and overjoyed to come to the ceremony.

What of the occasional Ig winners—those whose achievements might be said to be villainous or, well, stupid? *Those* achievements speak for themselves, all the more eloquently because they are in the company of other Ig-winning achievements that examplify whimsy, charm and self-deprecating humor.

The whole Ig ceremony is a study in contrasts, and a demonstration that most things in life are too rich and too ambiguous to take at face value.

The 1996 ceremony saw the inauguration of a new Ig ritual. Here, Lin Calista (center) of Cornucopia Auction Sales auctions off plaster casts of the left feet (on table at right) of Nobel Laureates Glashow, Herschbach, William Lipscomb and Richard Roberts. Scientist/supermodel Symmetra (right), who cast the feet, holds a plaster version of one of her own pedal extremities. Note the paper airplane (far left) approaching the stage. The auction led to a scandal, with two people each later claiming to have purchased Roberts's foot. Despite the Ig Nobel Committee's plea that the claimants "rectify this grievous misstep," the foot fiasco was written up in the Wall Street Journal. Photo: Enzo Crivelli/Mark Salza.

We *Are* Amused

The following editorial appeared in the October 7, 1996 issue of the British science journal
Chemistry & Industry. It is reprinted here with permission from *Chemistry & Industry*.

Is Britain's chief scientific adviser, Robert May, a pompous killjoy? In his recently publicised criticism of the Ig Nobel awards, a well-established spoof of the Nobel Prizes, he appears only to confirm that the British scientific establishment takes itself far too seriously.

In an interview with the journal *Nature,* May warns that the Ig Nobels risk bringing 'genuine' scientific projects into counter-productive ridicule. They should focus on anti-science and pseudo-science, he suggests, 'while leaving serious scientists to get on with their work.' His pique stems from embarrassing media coverage given to UK food scientists after an award last year for their research on soggy cereal flakes.

Such whining has several flaws. First, it is not for bureaucrats like May to determine which scientists are 'serious', or to ask that some researchers be ignored because they are above being made fun of (they aren't—the good ones as well as the bad ones).

Secondly, the Ig Nobels are organised by academics, for academics—unlike the notorious Golden Fleece awards in the US, with which May compares the Ig Nobels. The Ig Nobels let science laugh at itself.

Thirdly, the work of genuinely 'serious' scientists will withstand transitory embarrassment at the hands of TV comics and tabloid newspapers—assuming, of course, that their work really is recognised as 'serious' by other scientists. If, under a sudden spotlight, some scientists have to spend much time and effort explaining to everyone why their work is worth funding, that is a good thing and should happen more often, not less.

Finally, May reportedly suggests that the Ig Nobel organiser should obtain winners' consent first. But the British scientists *did* agree to receive their award last year, which makes May's grumbling distinctly off-target. Furthermore, that particular award proved that media mischief cannot be avoided by obtaining prior consent. As the Ig Nobel organiser, Marc Abrahams, has pointed out to May, 'there are few things, good or bad, that British tabloids and TV comedians do not ridicule.'

Far from making a convincing case for the pernicious effect of the Ig Nobels, May's misfire only makes him (and British science) look thin-skinned and humourless. He mistakes discomfort for disaster, and solemnity for seriousness. And he misunderstands the point, the process, and the pleasure of the awards. On this topic, scientists and others should reject this adviser's ill-advised views. Long may British scientists take their rightful places in the Ig Nobel honour roll.

The Ig Nobel Prizewinners

Here is a complete list of the Ig Nobel Prizewinners from 1996 through 1991, the year of the first ceremony. Several of the winners' acceptance speeches are presented here, too.

The audience prepares a typically genteel welcome for the 1995 Ig Nobel Prizewinners. Photo: Stephen Powell.

The 1996 Ig Nobel Laureates

Physics

Robert Matthews of Aston University, England, for his studies of Murphy's Law, and especially for demonstrating that toast often falls on the buttered side.

Chemistry

George Goble of Purdue University, for his blistering world record time for igniting a barbeque grill—three seconds, using charcoal and liquid oxygen.

Biology

Anders Baerheim and Hogne Sandvik of the University of Bergen, Norway, for their tasty and tasteful report, "Effect of Ale, Garlic, and Soured Cream on the Appetite of Leeches."

Medicine

James Johnston of R. J. Reynolds, Joseph Taddeo of U.S. Tobaccco, Andrew Tisch of Lorillard, William Campbell of Philip Morris, and the late Thomas E. Sandefur, Jr., chairman of Brown and Williamson Tobacco Co. for their unshakable discovery, as testified to the U.S. Congress, that nicotine is not addictive.

Literature

The editors of the journal *Social Text,* for eagerly publishing research that they could not understand, that the author said was meaningless, and which claimed that reality does not exist.

Economics

Dr. Robert J. Genco of the University of Buffalo for his discovery that "financial strain is a risk indicator for destructive periodontal disease."

Peace

Jacques Chirac, President of France, for commemorating the fiftieth anniversary of Hiroshima with atomic bomb tests in the Pacific.

Public Health

Ellen Kleist of Nuuk, Greenland and Harald Moi of Oslo, Norway, for their cautionary medical report "Transmission of Gonorrhoea Through an Inflatable Doll."

Biodiversity

Chonosuke Okamura of the Okamura Fossil Laboratory in Nagoya, Japan, for discovering the fossils of dinosaurs, horses, dragons, princesses, and more than 1000 other extinct "mini-species," each of which is less than 1/100 of an inch in length.

Art

Don Featherstone of Fitchburg, Massachusetts, for his ornamentally evolutionary invention, the plastic pink flamingo.

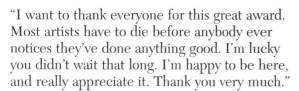

Ig Nobelliana
Words for the ages

"I want to thank everyone for this great award. Most artists have to die before anybody ever notices they've done anything good. I'm lucky you didn't wait that long. I'm happy to be here, and really appreciate it. Thank you very much."

— *Don Featherstone, winner of the 1996 Ig Nobel Art Prize for creating the plastic pink flamingo.*

ACCEPTANCE SPEECH

Robert Matthews
1996 Ig Nobel Physics Prize

Thank you very much for this award. Proving that Murphy's Law—if something can go wrong it will—is built into the design of the universe has brought me, as one of the most pessimistic people on earth, a lot of pleasure, and so has this Ig Nobel Prize. There is of course a more serious side to my work, I just can't remember what it is. Oh yes, I know. I should get out more.

ACCEPTANCE SPEECH

Dr. Harald Moi
1996 Ig Nobel Public Health Prize

Dr. Moi traveled from Oslo Norway to Harvard, at his own expense, to attend the ceremony. The following day, he delivered a more technical, and much longer, version of this lecture at Harvard Medical School.

Ladies and gentlemen, I am honored and delighted to accept the famous Ig Nobel Prize. I guess, however, I should not too honored, because this kind of research might easily be punctured.

The biggest problem in this case was how to perform the mandatory partner notification and treatment. No reference in the literature to the pharmakinetics of antibiotics in dolls could be found. So what else could be done than just give a shot and puncture?

1996 Art Prize winner Don Featherstone, the inventor of the plastic pink flamingo, delivers his acceptance speech at Harvard's Sanders Theatre. Featherstone was accompanied by his wife. The Featherstones wore matching bright pink suits. 1992 Art Prize winner Jim Knowlton, the creator of the classic anatomy poster, "Penises of the Animal Kingdom," is seated immediately behind the Featherstones. 1994 Entomology Prize winner Robert Lopez (ear mites), seated in the second row, watches Norwegian consul Terje Korsnes play with the Ig Nobel Prize that Korsnes accepted on behalf of Anders Baerheim and Hogne Sandvik. Note the flamingo neck visible just to the left of the podium and the paper airplanes littering the floor. Photo: Stephen Powell.

1996 Public Health co-Prizewinner Harald Moi giving his acceptance speech. Dr. Moi traveled from Oslo, Norway to Cambridge, at his own expense, to receive the Prize for his report "Transmission of Gonorrhoea Through an Inflatable Doll." Photo: Stephen Powell.

The 1995 Ig Nobel Laureates

Physics

D. M. R. Georget, R. Parker, and A. C. Smith, of the Institute of Food Research, Norwich, England, for their rigorous analysis of soggy breakfast cereal, published in the report entitled "A Study of the Effects of Water Content on the Compaction Behaviour of Breakfast Cereal Flakes."

Chemistry

Bijan Pakzad of Beverly Hills, for creating DNA Cologne and DNA Perfume, neither of which contain deoxyribonucleic acid, and both of which come in a triple helix bottle.

Medicine

Marcia E. Buebel, David S. Shannahoff-Khalsa, and Michael R. Boyle, for their invigorating study enti-

tled "The Effects of Unilateral Forced Nostril Breathing on Cognition."

Literature

David B. Busch and James R. Starling, of Madison, Wisconsin, for their deeply penetrating research report, "Rectal Foreign Bodies: Case Reports and a Comprehensive Review of the World's Literature." The citations include reports of, among other items: seven light bulbs; a knife sharpener; two flashlights; a wire spring; a snuff box; an oil can with potato stopper; eleven different forms of fruits, vegetables and other foodstuffs; a jeweler's saw; a frozen pig's tail; a tin cup; a beer glass; and one patient's remarkable ensemble collection consisting of spectacles, a suitcase key, a tobacco pouch and a magazine.

Economics

Awarded jointly to Nick Leeson and his superiors at Barings Bank and to Robert Citron of Orange County, California, for using the calculus of derivatives to demonstrate that every financial institution has its limits.

Peace

The Taiwan National Parliament, for demonstrating that politicians gain more by punching, kicking and gouging each other than by waging war against other nations.

Psychology

Shigeru Watanabe, Junko Sakamoto, and Masumi Wakita, of Keio University, for their success in training pigeons to discriminate between the paintings of Picasso and those of Monet.

Nutrition

John Martinez of J. Martinez & Company in Atlanta, for Luak Coffee, the world's most expensive coffee, which is made from coffee beans ingested and excreted by the luak (aka, the palm civet), a bobcat-like animal native to Indonesia.

Public Health

Martha Kold Bakkevig of Sintef Unimed in Trondheim, Norway, and Ruth Nielson of the Technical University of Denmark, for their exhaustive study,

At the 1995 ceremony, The Nicola Hawkins Dancers serve up a surprise for the five Nobel Laureates—steaming hot mugs of Luak Coffee, brewed from coffee beans that were eaten and excreted by luaks. Luaks are bobcat-like animals native to Indonesia. Photo: Alexandra Murphy.

"Impact of Wet Underwear on Thermoregulatory Responses and Thermal Comfort in the Cold."

Dentistry

Robert H. Beaumont, of Shore View, Minnesota, for his incisive study "Patient Preference for Waxed or Unwaxed Dental Floss."

ACCEPTANCE SPEECH

Andrew Smith
Institute of Food Research,
Norwich, England
1995 Ig Nobel Physics Prize

In our study of compaction of breakfast cereal flakes, we did not leave them turned tongue-in-cheek, or use any other sensory technique. Rather, we set out to relate macro-scale mechanical properties to changes in the scale of constituent food particle molecules. This provides valuable insights into texture. So what does this mean for the manufacturer, and to you, the consumer? Well, it's all about the quest for the ultimate breakfast cereal-eating experience. I hope that the awarding of this prize will stimulate further research in this area.

ACCEPTANCE SPEECH

Sally Yeh
President, Bijan Fragrances, Inc.,
Beverly Hills, California
1995 Ig Nobel Chemistry Prize

Ms. Yeh delivered this acceptance speech on behalf of Bijan Pakzad.

It is my honor and pleasure to accept this award on behalf of the world-renowned, most exclusive menswear and fragrance designer, Bijan. Now, Bijan is unable to attend, but he is very excited about this event. Right now he is filming his new advertising campaign with the famous actress Bo Derek. Together, they should make a perfect twenty. Undeniably humbled by the presence of such notable personalities in this room, I have only this to say about DNA. In our mind, DNA not only stands for "deoxyribonucleic acid"—it also captures the initials of Bijan's three children, Daniela, Nicolas and Alexandra. They

Sally Yeh, president of Bijan Fragrances, delivers an impassioned acceptance speech on behalf of 1995 Ig Nobel Chemistry Prize winner Bijan Pakzad, the creator of DNA Cologne and DNA Perfume. Photo: LeeAnn Tzeng.

Herschbach (left, reaching down for a bucket), Glashow, Murray and Lipscomb sample Luak Coffee. Roberts (not pictured here) also indulged. Photo: Jeff Pietrantoni.

are the only ones who don't have to go to department stores to get designer jeans. Actually, when Bijan created these award-winning fragrances he had in mind for DNA to be interpreted as "Definitely Not Average" and "Damn Near Affordable." So on behalf of the designer Bijan and everyone at Bijan Fragrances, I thank you for this award.

The 1994 Ig Nobel Laureates

Physics

The Japanese Meterological Agency, for its seven-year study of whether earthquakes are caused by catfish wiggling their tails.

Chemistry

Texas State Senator Bob Glasgow, wise writer of logical legislation, for sponsoring the 1989 drug control law which makes it illegal to purchase beakers, flasks, test tubes, or other laboratory glassware without a permit.

Biology

W. Brian Sweeney, Brian Krafte-Jacobs, Jeffrey W. Britton, and Wayne Hansen, for their breakthrough study, "The Constipated Serviceman: Prevalence Among Deployed U.S. Troops," and especially for their numerical analysis of bowel movement frequency.

Medicine

This prize is awarded in two parts. First, to Patient X, formerly of the U.S. Marine Corps, valiant victim of a venomous bite from his pet rattlesnake, for his determined use of electroshock therapy—at his own insistence, automobile sparkplug wires were attached to his lip, and the car engine revved to 3000 rpm for five minutes. Second, to Dr. Richard C. Dart of the Rocky Mountain Poison Center and Dr. Richard A. Gustafson of The University of Arizona Health Sciences Center, for their well-grounded medical report, "Failure of Electric Shock Treatment for Rattlesnake Envenomation."

Literature

L. Ron Hubbard, ardent author of science fiction and founding father of Scientology, for his crackling Good Book, *Dianetics*, which is highly profitable to mankind or to a portion thereof.

Economics

Jan Pablo Davila of Chile, tireless trader of financial futures and former employee of the state-owned Codelco Company, for instructing his computer to "buy" when he meant "sell," and subsequently attempting to recoup his losses by making increasingly unprofitable trades that ultimately lost .5 percent of Chile's gross national product. Davila's relentless achievement inspired his countrymen to coin a new verb: "davilar," meaning, "to botch things up royally."

Peace

John Hagelin of Maharishi University and The Institute of Science, Technology and Public Policy, promulgator of peaceful thoughts, for his experimental conclusion that 4,000 trained meditators caused an 18 percent decrease in violent crime in Washington, D.C.

Mathematics

The Southern Baptist Church of Alabama, mathematical measurers of morality, for their county-by-county estimate of how many Alabama citizens will go to Hell if they don't repent.

Entomology

Robert A. Lopez of Westport, N.Y., valiant veterinarian and friend of all creatures great and small, for his series of experiments in obtaining ear mites from cats, inserting them into his own ear, and carefully observing and analyzing the results.

Psychology

Lee Kuan Yew, former Prime Minister of Singapore, practitioner of the psychology of negative reinforcement, for his thirty-year study of the effects of punishing three million citizens of Singapore whenever they spat, chewed gum, or fed pigeons.

ACCEPTANCE SPEECH

Tim Mitchell
Science Products Division, Corning, Inc.
1994 Ig Nobel Chemistry Prize

Mr. Mitchell accepted custody of the Prize on behalf of the winner, Texas State Senator Bob Glasgow.

I am here to accept this in lieu of the actual winner. Tonight I'll use this forum to make a few comments on a hot social and scientific issue brought to light by the lawmakers in Texas. That topic is the unregulated and unrestricted sale of test tubes, beakers, and other laboratory apparatus in America. There is a grassroots movement out there to convince the state of Texas to amend their laboratory glassware law. Instead of outlawing glassware altogether, this group would like to see a five day cooling-off period. They feel this will be enough to discourage people from purchasing a beaker and then using it in a fit of rage to harm themselves or others. Part of me wonders, will a waiting period be enough? You see, it only starts with a test tube. You think to yourself, hey, it's only a test tube, for God's sake. Pretty soon, though, the rush from a test tube isn't enough. You want to experiment more and more. Then before you know it, you're laying in the corner of a lab somewhere with a soxilette apparatus in one hand, a three neck flask in the other, strung out and begging for grant money.

ACCEPTANCE SPEECH

Dr. Richard Dart
Director, Rocky Mountain Poison Center
1994 Ig Nobel Medicine Prize

I was stunned to receive this prize, although not as stunned as our patient.

Dr. Richard Dart, co-winner of the 1994 Medicine Prize. Dart and a colleague used conventional therapy to treat the snakebit, self-electroshocked Patient X.

ACCEPTANCE SPEECH

Terje Korsnes
Honorary Consul from Norway
   to Massachusetts
1994 Ig Nobel Mathematics Prize

Mr. Korsnes accepted custody of the Prize on behalf of the Southern Baptist Church of Alabama, who won the Prize, and on behalf of the citizens of the town of Hell, Norway.

I was asked to come here tonight and accept custody of this prize on behalf of the people of Hell, Norway. We were delighted to learn that so many people in the great state of Alabama will go to Hell. We have a special place in Hell for all of you.

I have a few of these guys with me, and I want to pass them out to the audience. Thank you very much.

Einstein's Dreams author and MIT astronomer Alan Lightman delivers a tribute to Ig 1994 Nobel Literature Prize winner L. Ron Hubbard. Despite expectations, Hubbard failed to appear on stage. As Lightman described Hubbard's career, a balloon figure of unknown origin floated toward the stage and lazily circled the podium. The balloon was destroyed before it could cause any harm.

ACCEPTANCE SPEECH

Dr. Robert Lopez
1994 Ig Nobel Entomology Prize

> I hate the old didactic mite.
> All he does is crawl around and bite.
> At sleeping time he acts just like a bum,
> Crawls right into your ear drum.
> Once there, he scratches and bites *ad infinitum*.

The 1993 Ig Nobel Laureates

Physics

Louis Kervran of France, ardent admirer of alchemy, for his conclusion that the calcium in chickens' eggshells is created by a process of cold fusion.

Chemistry

James Campbell and Gaines Campbell of Lookout Mountain, Tennessee, dedicated deliverers of fragrance, for inventing scent strips, the odious method by which perfume is applied to magazine pages.

Biology

Paul Williams Jr. of the Oregon State Health Division and Kenneth W. Newell of the Liverpool School of Tropical Medicine, bold biological detectives, for their pioneering study, "Salmonella Excretion in Joy-Riding Pigs."

Medicine

James F. Nolan, Thomas J. Stillwell, and John P. Sands, Jr., medical men of mercy, for their painstaking research report, "Acute Management of the Zipper-Entrapped Penis."

Literature

Awarded jointly to E. Topol, R. Califf, F. Van de Werf, P. W. Armstrong, and their 972 co-authors, for publishing a medical research paper which has one hundred times as many authors as pages.

Economics

Ravi Batra of Southern Methodist University, shrewd economist and best-selling author of *The Great Depression of 1990* ($17.95) and *Surviving the Great Depression of 1990* ($18.95), for selling enough copies of his books to single-handedly prevent worldwide economic collapse.

Peace

The Pepsi-Cola Company of the Phillipines, suppliers of sugary hopes and dreams, for sponsoring a contest to create a millionaire, and then announcing the wrong winning number, thereby inciting and uniting 800,000 riotously expectant winners, and bringing many warring factions together for the first time in their nation's history.

Mathematics

Robert Faid of Greenville, South Carolina, far-sighted and faithful seer of statistics, for calculating the exact odds (8,606,091,751,882:1) that Mikhail Gorbachev is the Antichrist.

Visionary Technology

Presented jointly to Jay Schiffman of Farmington Hills, Michigan, crack inventor of AutoVision, an image projection device that makes it possible to drive a car and watch television at the same time, and to the Michigan state legislature, for making it legal to do so.

Psychology

John Mack of Harvard Medical School and David Jacobs of Temple University, mental visionaries, for their leaping conclusion that people who believe they were kidnapped by aliens from outer space, probably were—and especially for their conclusion that "the focus of the abduction is the production of children."

Consumer Engineering

Ron Popeil, incessant inventor and perpetual pitchman of late night television, for redefining the industrial revolution with such devices as the Veg-O-Matic, the Pocket Fisherman, the Cap Snaffler, Mr. Microphone, and the Inside-the-Shell Egg Scrambler.

1993 Medicine Prize co-winner Dr. James Nolan made a twelve hour drive to explain his role in the painstaking clinical report, "Acute Management of the Zipper-Entrapped Penis." After Nolan's speech, the entire sellout crowd of 1200 rose and serenaded him with a modified version of the Michael Jackson song "We Are the World." Photo: Roland Sharrillo.

penile predicaments. My colleagues and I at the Navy Hospital in San Diego (where we performed the research) and a competing group at the University of California, San Francisco, have already shed more light on the management of the human bite to the penis—a timely topic for the Navy due to the upcoming changes promulgated by the present administration. And now, as I change my career path to be a urologist in rural America, my colleagues and I at the Guthrie Clinic hope to further clarify the incidence and significance of urologic trauma secondary to farm animals.

ACCEPTANCE SPEECH

Dr. James Nolan
1993 Ig Nobel Medicine Prize

I wish my mother was here to see me accept this prize. My colleagues and I never dreamed this simple paper would attract so much attention. I was here to save my generation from penile injury. Your recognition here tonight has stimulated my interest in further pursuing research in the field of painful

ACCEPTANCE SPEECH

Kevin Steiling
Assistant Attorney General,
    Commonwealth of Massachusetts
1993 Ig Nobel Psychology Prize

Mr. Steiling accepted the Prize on behalf of John Mack and David Jacobs.

My name is Kevin Steiling. I am Assistant Attorney General for the Commonwealth of Massachusetts.

Kidnapping is a federal offense. It is also a criminal act under the statutes of the Commonwealth of Massachusetts. Last year there were hundreds of kidnappings or attempted kidnappings. None of them involved aliens from other planets. Thank you.

After the ceremony, John Mack's assistant telephoned the Ig Nobel Board of governors to request that Dr. Mack be invited to deliver a speech at the next year's ceremony. The following year, elaborate preparations were made, but at the last moment Dr. Mack indicated that he would not, after all, attend to the ceremony. At that second ceremony, the Ig Nobel Board of Governors announced that "we are disappointed and hurt, yes, but above all we are concerned."

The 1992 Ig Nobel Laureates

Physics

David Chorley and Doug Bower, lions of low-energy physics, for their circular contributions to field theory based on the geometrical destruction of English crops.

Chemistry

Ivette Bassa, constructor of colorful colloids, for her role in the crowning achievement of twentieth century chemistry, the synthesis of bright blue Jell-O.

Biology

Dr. Cecil Jacobson, relentlessly generous sperm donor, and prolific patriarch of sperm banking, for devising a simple, single-handed method of quality control.

Medicine

F. Kanda, E. Yagi, M. Fukuda, K. Nakajima, T. Ohta and O. Nakata of the Shisedo Research Center in Yokohama, for their pioneering research study, "Elucidation of Chemical Compounds Responsible for Foot Malodour," and especially for their conclusion that people who think they have foot odor do, and those who don't, don't.

Literature

Yuri Struchkov, unstoppable author from the Institute of Organoelemental Compounds in Moscow, for the 948 scientific papers he published between the years 1981 and 1990, averaging more than one every 3.9 days.

Economics

The investors of Lloyds of London, heirs to 300 years of dull prudent management, for their bold attempt to insure disaster by refusing to pay for their company's losses.

Peace

Daryl Gates, former Police Chief of the City of Los Angeles, for his uniquely compelling methods of bringing people together.

Nutrition

The utilizers of Spam, courageous consumers of canned comestibles, for 54 years of undiscriminating digestion.

Archeology

Eclaireurs de France, the Protestant youth group whose name means "those who show the way," fresh-scrubbed removers of grafitti, for erasing the ancient paintings from the walls of the Meyrieres Cave near the French village of Brunquiel.

Art

Presented jointly to Jim Knowlton, modern Renaissance man, for his classic anatomy poster "Penises of the Animal Kingdom," and to the U.S. National Endowment for the Arts for encouraging Mr. Knowlton to extend his work in the form of a pop-up book.

The 1991 Ig Nobel Laureates

Physics

(*) Thomas Kyle, detector of atoms and original man of knowledge, for his discovery of the heaviest element in the universe, Administratium.

(*) At the first Ig Nobel Prize Ceremony in 1991, three of the seven Prizes were given to individuals who may have been fictional, fraudulent or indeterminate. The other winners that year—*and all winners in subsequent years*—have been and are completely genuine.

An Official Heckler Corps leads the 1994 audience members in serene intellectual discourse.

Chemistry

Jacques Benveniste, prolific proseletizer and dedicated correspondent of *Nature*, for his persistent discovery that water, H_2O, is an intelligent liquid, and for demonstrating to his satisfaction that water is able to remember events long after all trace of those events has vanished.

Biology

Robert Klark Graham, selector of seeds and prophet of propagation, for his pioneering development of the Repository for Germinal Choice, a sperm bank that accepts donations only from Nobellians and Olympians.

Medicine

Alan Kligerman, deviser of digestive deliverance, vanquisher of vapor, and inventor of Beano, for his pioneering work with anti-gas liquids that prevent bloat, gassiness, discomfort and embarrassment.

Literature

Erich Von Daniken, visionary raconteur and author of *Chariots of the Gods*, for explaining how human civilization was influenced by ancient astronauts from outer space.

Economics

Michael Milken, titan of Wall Street and father of the junk bond, to whom the world is indebted.

Peace

Edward Teller, father of the hydrogen bomb and first champion of the Star Wars weapons system, for his lifelong efforts to change the meaning of peace as we know it.

Interdisciplinary Research

(*) Josiah Carberry of Brown University, bold explorer and eclectic seeker of knowledge, for his pioneering work in the field of Psychoceramics, the study of cracked pots.

Education

J. Danforth Quayle, consumer of time and occupier of space, for demonstrating, better than anyone else, the need for science education.

Pedestrian Technology

(*) Paul DeFanti, wizard of structures and crusader for public safety, for his invention of the Buckybonnet, a geodesic fashion structure that pedestrians wear to protect their heads and preserve their composure.

Members of the Harvard-Radcliffe Science Fiction Association make eye contact with unidentified beings at the 1996 ceremony. Photo: Enzo Crivelli/Mark Salza.

Those Who Covet the Ig

How Washington lobbyists and others tried to hijack the Ig Nobel Prizes

by Stephen Drew

The Ig Nobel Prizes inspire strong emotions. Some winners are delighted to receive them, others less so. But the Ig Nobel Board of Governors was surprised to find that some people yearn not to receive Ig Nobel Prizes, but to give them.

A Washington-based animal rights lobby group attempted, and failed, to appropriate the good name of the Ig Nobel Prizes. On December 28, 1994, a group calling itself "Physicians' Committee for Responsible Medicine" issued a press release in which it tried to announce its own list of Ig Nobel Prize-winners. The international science community was shocked.

"I am shocked," said Harvard Professor William Lipscomb, a 1976 Nobel Laureate in Chemistry. "I am shocked," said Harvard Professor Sheldon Glashow, a 1990 Nobel Laureate in Physics. "It's outrageous. My hair stands on end at the very thought of it," said New England Biolabs research director Richard Roberts, a 1993 Nobel Laureate in Physiology or Medicine. "I am shocked and disgusted," said Harvard Professor Dudley Herschbach, a 1979 Nobel Laureate in Chemistry.

"I am appalled that someone would try to use the vehicle of the Ig Nobel awards for political aims," said MIT Professor Jerome Friedman, a 1990 Nobel Laureate in Physics. "The purpose of these awards is to enhance the humor of our lives, something that is in short supply and should be protected."

The Ig Nobel Board of Governors issued the following statement:

> We were shocked, *shocked* to hear that a lobbying group in Washington, D.C. has tried to claim credit for the Ig Nobel Prizes. Maybe they're just frustrated that they have never won an Ig Nobel Prize and feel this is a novel way to campaign for one.

The matter reached the news media, where it was the subject of much amusement. Then on January 9, 1995, after receiving phone calls, faxes and letters from outraged citizens of many nations, Neal D. Barnard, president of the lobbying group, sent a letter to *AIR*. The letter reads:

> Please be assured that we had no intention of intentionally [sic] appropriating the name of your annual award. I regret any confusion we may have caused.

Two years later, a man in Illinois proclaimed that he would be giving out an Ig Nobel Prize. The Ig Nobel Board of Governors thereupon issued a public request:

> If you hear of any other groups or individuals that plan to give out Ig Nobel Prizes, please drop us a line. We will then invite all these people to a party (organized at their own expense) at which they can gather and chew both the fat and each other.

Transmission of Gonorrhoea Through an Inflatable Doll

by E. Kleist
Nanortalik Hospital
Nanortalik, Greenland
and H. Moi
Post Box 1001, Venereaklinikken
3900 Nuuk, Greenland

Ellen Kleist and Harald Moi won the 1996 Ig Nobel Prize in Public Health for this letter that appeared in *Genitourinary Medicine*, vol. 69, 1993, p. 322. It is reprinted here with permission. The references are as they appeared in the original.

Nonsexual transmission of gonorrhoea seems to be extremely rare. Only one case of nonsexual transmission of genital Neisseria gonorrhoeae is documented in adults,[1] involving two patients in a military hospital who shared a urinal. N. gonorrhoeae has been shown to survive in infected secretions on towels and handkerchiefs for 20 and 24 hours, respectively.[2] Cultures from toilet seats in public restrooms and venereal disease clinics have failed to yield N. gonorrhoeae.[3,4]

The skipper from a trawler, who had been 3 months at sea, sought advice for urethral discharge. His symptoms had lasted for two weeks. A urethral smear showed typical intracellular gram-negative diplococci, and a culture was positive for N. gonorrhoeae. There had been no woman on board the trawler; he denied homosexual contacts; and there was no doubt that the onset of the symptoms was more than two months after leaving the port.

With some hesitation, he told the story. A few days before onset of his symptoms, he had roused the engineer in his cabin during the night because of engine trouble. After the engineer had left his cabin, the skipper found an inflatable doll with artificial vagina in his bed, and he was tempted to have "intercourse" with the doll. His complaints started a few days after this episode.

The engineer was examined, and was found to have gonorrhoea. He had observed a mild urethral discharge since they left port, but he had not been treated with antibiotics. He admitted to having ejaculated into the "vagina" of the doll just before the skipper called him, without washing the doll afterwards. He also admitted intercourse with a girl in another town some days before going to sea. This girl was traced, but the result of her examination is not known. To the best of our knowledge, no case of gonococcal transmission through an inflatable doll had been reported before.

Notes

1. Neinstein, L. S., Goldenring J., Carpenter, S. Nonsexual transmission of sexually transmitted diseases: an infrequent occurrence. *Pediatrics* 1984; 74:67–76.
2. Srivastava, A. C. Survival of gonococci in urethral secretions with reference to the nonsexual transmission of gonococcal infection. *J. Med. Microbiol.* 1980; 13:593–6.
3. Gilbaugh, J. H., Juchs, P.C. The gonococcus and the toilet seat. *N. Engl. J. Med.* 1979; 301:91–3.
4. Rein, M. F. Nonsexual acquisition of gonococcal infection (letter). *N. Engl. J. Med.* 1979; 301:1347.

Of Mites and Man

by Robert A. Lopez, DVM
Westport, New York

Robert Lopez won the 1994 Ig Nobel Prize in Entomology for this letter that appeared in *Journal of the American Veterinary Medical Association*, vol. 203, no. 5, 1993, pp. 606–7. It is reprinted here with permission. The reference at the end is as it appears in the original.

Two strange and related clinical cases prompted me to investigate the possibility of transmission of the ear mite, *Otodectes cynotis,* to human beings. In the first case, a client, accompanied by her three-year-old daughter, brought in two cats with severe ear mite infestations. In the examining room, the daughter happened to complain of itching chest and abdomen. The mother stated that the daughter frequently held the cats for long periods, like dolls, then showed me the numerous small red abdominal bite marks that were the source of the itching. I recommended that she check with her pediatrician. After the cat's ear mite infestation was cleared up, I learned that the daughter's itching also quickly disappeared.

A year later, when the same client brought in a cat heavily infested with ear mites, she complained of bites on her ankles. The bites subsequently stopped after the cat had been cleared of the ear mite infestation.

At that time (1968), a search of the literature did not reveal any report of *Otodectes cynotis* infestation in human beings, so I decided to be a human guinea pig.

I obtained ear mites from a cat and confirmed by microscopic examination that they were *Otodectes cynotis.* Then I moistened a sterile cotton-tipped swab with warm tap water and transferred approximately one g of ear mite exudate from the cat to my left ear. Immediately, I heard scratching sounds, then moving sounds, as the mites began to explore my ear canal. Itching sensations then started, and all three sensations merged into a weird cacophony of sound and pain that intensified from that moment, 4 PM, on and on. . . . At first, I thought this wouldn't and couldn't last very long. However, as the day and evening wore on, I began to worry. The pruritus was increasing. The sounds in my ear (fortunately I had chosen only one ear), were becoming louder as the mites traveled deeper toward my ear drum. I felt helpless. Is this the way a mite-infested animal feels?

For the next five hours the mites were very active and then their activity, measured by scratching sounds and degree of pruritus, leveled off. There still was something definitely crawling about deep in my left ear, but the discomfort was bearable.

After retiring about 11 PM, the mite activity increased incrementally so that, by midnight, the mites were very busy, biting, scratching, and moving about. By 1 AM, the sounds were loud. An hour later the pruritus was very intense. After two hours the highest level of itching and scratching was attained. Sleep was impossible. Then, suddenly, the mites seemed to lessen their feeding activities. The noise and pruritus abated, and a brief sleep was possible. Mite activity resumed at 7 AM, with light noises and slight pruritus. This pattern was repeated—light mite activity during the day, with slight increase in the evening, from approximately 6 to 9 PM, and then heavy mite activity from midnight to 3 AM. This night feeding pattern was quite regular and made sleep, no matter how demanding, completely out of the question.

By the second week, when the late night feeding pattern had become well established, the intensity of the mite activity began to lessen. By the third week, the ear canal was filling up with debris, and hearing from my left ear was gone. By the fourth week, mite activity was 75% reduced and I could feel mites crawling across my face at night. They never did try to enter my right ear, nor did they bite or cause any itching anywhere else on my body. At the end of one month, I could no longer hear or feel any mite activity. The pruritus and internal ear noises were going. However, my ear was completely filled with exudate. I cleansed my ear with warm water swabs and flushings, for the first time. Within one week, my left ear was clear of debris. By the sixth week, there was no pruritus, and hearing was normal. Recovery was surprisingly fast with just warm water irrigations.

By the eighth week, I decided to try again to see whether the first experiment had been flawed or misleading. Accordingly, with my left ear now healed, no debris evident, and hearing normal, I obtained ear mites from another cat and confirmed their identity as *Otodectes cynotis,* as I had done before. I transferred a 1- to 2-g sample of the cat's ear exudate to my left ear, as I had done before. Once again, my ear began to react to the mite invasion. Loud scratching noises, pruritus, and pain all began within a few seconds. The same pattern evolved, with an early evening feeding pattern and a late heavy eating session. Intermittent feeding forays during the day were short. The first week was again filled with intense pruritus, and the second week with lessened mite activity that ceased by day 14. The left ear was filled with much less exudate, and my hearing was only slightly impaired. Warm water irrigations cleared up the infestation in 72 hours.

This definite reduction in symptoms left many questions. Was there an immunity? Were human ears refractory to *Otodectes?* A third and final trial had to be done.

At week 11, I repeated the experiment as before, using my left ear. Within a few minutes, the pruritus and inner ear noises began. However, this time the intensity was much less severe. Very little debris accumulated, and hearing was only partially affected. Feeding patterns remained the same. By the end of day eight or nine, the mites had ceased biting, although I had felt some walking across my face during the night. Once again, nothing but warm water was used to rinse the left ear. It healed again quickly, except for occasional slight bouts of pruritus.

This descending time and intenstiy of infestation or increasing immunity under similar modes of experimentation raises some interesting questions. Is there an immune reaction to parasitisms, particularly *Otodectes cynotis,* in mammals? In over 30 years of practice, I have noticed that the younger cats had more severe ear mite infestations.

Do otodectic mites have a regular feeding pattern? If so, would late evening treatments be more effective? I routinely advise clients to use ear medications for mites late in the evenings.

Since my initial literature search, I have found one report of natural ear mite infestation in a human being,[1] causing tinnitus. I wonder whether the person involved enjoyed her experience as much as I did.

Note

1. Suetake, M., Yuasa, R., Saijo, S., et al. Canine ear mites *Otodectes cynotis* found on both tympanic membranes of adult woman causing tinnitus. *Tohoku Rosat Hosp Proact Otolog Kyoto* 1991; 84:38–42.

Failure of Electric Shock Treatment for Rattlesnake Envenomation

by Richard C. Dart, MD and Richard A. Gustavson

Westport, New York
University of Arizona Health Sciences Center
Tucson, Arizona

Richard Dart, Richard Gustavson, and "Patient X" won the 1994 Ig Nobel Prize in medicine for this report that appeared in *Annals of Emergency Medicine*, vol. 20, no. 6, 1991, pp. 659–61. It is reprinted here with permission. This is a shortened version. The references are as they appeared in the original.

Introduction

The application of high-voltage electric shock therapy, low-current electric shock treatment has been reported as an effective treatment for rattlesnake and other envenomations. This concept was proposed in a letter describing dramatic improvement of pain and swelling in victims of snakebite in a remote region of Ecuador.[1] The initial report was followed by several articles in medical journals and the popular press,[2–5] resulting in the widespread adoption of this concept and in the sale of portable, self-operated electric shock devices for use on animal and human victims of snakebite.[6,7]

There have been few reports documenting the use of electric shock for treatment of envenomation of human beings. Proponents of electric shock therapy cite anecdotal reports asserting the benefits to human beings in support of its use, despite its failures in animal models. We report a case demonstrating the potential danger and the ineffectiveness for antivenin administration in patients with a history of sensitivity to antivenin are presented.

Case Report

The Arizona Poison and Drug Information Center was consulted concerning a 28-year-old man who had been handling his pet Great Basin rattlesnake (*Crotalus viridis lutosus*) when he was bitten near the right upper lip. The patient had a history of 14 prior bites and had previously developed anaphylactic shock secondary to antivenin administration. The patient noted burning and swelling of the face within minutes of the bite. Based on their understanding of an article in an outdoorsman's magazine, the patient and his neighbor had previously established a plan to use electric shock treatment if either was envenomated. The patient was placed supine next to a car, and a spark plug wire was attached to his upper lip by a wire with a small clip at each end. The engine was started and repeatedly revved to 3,000 rpm for approximately five minutes. The patient lost consciousness with the first electrical charge.

An ambulance arrived approximately 15 minutes later to find the patient unconscious and incontinent of stool. Initial vital signs were blood pressure of 100 mm Hg and palpable; pulse, 100; and respirations, 20. He was transported to an emergency department by helicopter. The patient became combative when nasal intubation was attempted during transport. Blood pressure was reported as in the 80s.

On arrival at the ED one hour and 40 minutes after the bite, the patient remained obtunded with blood pressure of 62 mm Hg and palpable; pulse, 120; and temperature, 35.4 C. Severe facial edema

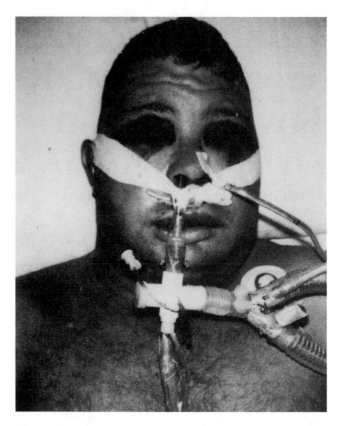

and accymosis were noted. The patient was paralyzed and nasotracheally intubated because of increasing facial, neck, and laryngeal swelling (Figure 1.) . . .

A skin test to antivenin (CROTALIDAE) polyvalent resulted in a large erythematous wheal. The patient received hydrocortisone 200 mg IV, diphenhydramine 100 mg IV, and cefazolin 1 g IV before the administration of antivenin. Antivenin was infused through a central line at an initial rate of 24 mL/hr. . . .

The patient was later extubated without difficulty. After four days of hospitalization, he developed serum sickness. At discharge, the patient had some facial edema and tissue loss of his upper lip. He has subsequently undergone reconstructive surgery of the lip.

Summary

Although highly touted for treatment of pit viper, arthropod, and hymenoptera envenomations, there is no evidence to support the use of electric shock. Anecdotal reports of its success in the Ecuadorian jungle continue but remain unpublished.[8] Because animal studies have not shown a beneficial effect, it is strongly advised that electric shock therapy not be used for the treatment of pit viper poisonings in the United States.

Figure 1

References

1. Guderian, R. H, Mackenzie, C. D, Williams, J. F: High voltage shock treatment for snakebite. *Lancet* 1986; 2:229.
2. High-voltate shock treatment for snakebite. *Postgrad. Med.* 1987; 82:250
3. Herzberg, R.: Shocks for snakebites. *Outdoor Life* 1987; June: 55–57, 110.
4. Mueller, L.; A shock cure. *Outdoor Life* 1988; June; 64–65,110–112.
5. Mueller, L.; A shock cure. *Outdoor Life* 1988; 45–47, 76–78.
6. Canine snakebite kit, in *Master Vaccine Catalog.* Spring 1988, p. 20.
7. Snake Doctor, promotional brochure. Claremore, Oklahoma, J and K Industries.
8. Hardy, D. L: Appropriate first aid measures for venemous snakebite should not come as a shock. *Tucson Herpetological Soc. Newsletter* 1988; 1:12–13.

The Okamura Fossil Laboratory

by Earle E. Spamer
Academy of Natural Sciences
Philadelphia, Pennsylvania

Chonosuke Okamura won the 1996 Ig Nobel Prize in the field of Biodiversity. This appreciation of Okamura's work appeared in *AIR* 1:4 in July/Aug 1994. The author, Earle Spamer, is responsible for bringing Okamura's work to the attention of a soon-to-be-adoring public. Spamer presented a brief version of this work as part of the 1996 Ig Nobel Prize Ceremony.

Is Chonosuke Okamura still alive and working? We do not know, but fear the worst. Earle Spamer recently abandoned a three-year quest to find Okamura. Whatever, the man's many small contributions to knowledge will live forever.

Okamura and the mini-man

The cover of *AIR* 2:4 (July/August 1995) featured both Chonosuke Okamura and an example of his work. The accompanying photomicrograph first appeared on page 527 of the *Original Report of the Okamura Fossil Laboratory*, where it was described thusly:

> In the picture, a [sic] apparently is an old man. . . . It is evident from the picture that the skin of the aged has become rough because of losing elasticity wich [sic] that of the youth is smooth and lively. . . . The old man has his mouth shut tight on the stretch, showing his character as a result of a long-year's mental activity since his younger.

Annals of Improbable Research

The Journal of Record for Inflated Research and Personalities

Vol. I, No.4 July/August 1995 $4.95

Okamura and the Minicreatures (see page 4)

Cindy Crawford, Woodchucks Tornadoes & Trailer Parks James Watson Interview

and free pheromones!!

The halls of science are emblazoned with portraits of genius. They are solitary memorials to the fruits of inspiration, reminders of how the works of individuals have changed humankind's view of the universe. Each field of study can fan the flames of scientific patriotism with its own scroll of saluted intellectuals. Unfortunately, the works of others go unnoticed, or underappreciated, during their lives.

In Nagoya, Japan, the 1970s and 1980s bore witness to the discovery and study of some of the most remarkable, yet least noticed, fossil vertebrates. The scholarly attention to them by one man, Chonosuke Okamura, has done no less than shake the foundations of paleontology, anthropology, and archaeology. In meticulously documented reports, Okamura has pushed back into the early Paleozoic Era the evidence for the evolution of the vertebrates and the rise of civilization, having discovered the earliest ancestors of modern vertebrate animals, including man.

Okamura's studies have all been of the Paleozoic limestones of Japan. Cutting and polishing smooth surfaces on his samples, Okamura examined the surfaces under the microscope, whereafter he was able to document the existence of a multitude of very small-scale fossils. His first studies were transmitted to meetings of the Paleontological Society of Japan (PSJ). He formally published his findings in the first of the series, *Original Report of the Okamura Fossil Laboratory (OROFL)*; in it he describes invertebrate and algal fossils ranging in age from the Ordovician Period to the Tertiary Period. He sees that the evolution of these organisms was rapid, quickly becoming similar to modern forms even during the Paleozoic.

The *OROFL* series is published in English, with Japanese abstracts, and profusely illustrated with photomicrographs, some in color. New fossil taxa are eloquently described, and for each, every aspect of the fossil is illustrated.

That Okamura's discoveries were not noticed by the scientific community is not for their having been disqualified by other evidence, but more likely because of the limited distribution of *OROFL*. Not being in "mainstream" or other "recognized" literature, Okamura's findings have been overlooked by the very researchers who would have benefited from these new data from the fossil record. It thus falls upon later workers to disseminate Okamura's results anew, in a more widely subscribed medium.

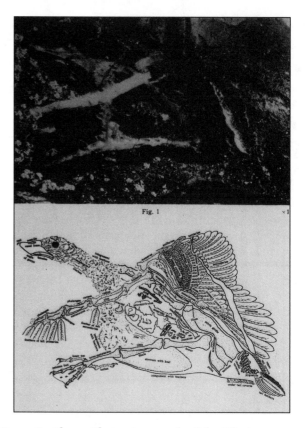

Figure 1: *Above: Photomicrograph of the Silurian miniduck, Archaeoanus japonica (OROFL 13, plate 31). Below: Okamura's diagram of same.*

Birds of a Feather

Until the mid-1970s, Okamura's microscopical studies were limited to the invertebrate and algal fossils. Then at the PSJ meeting of 18 June 1977, published two months later in *OROFL* 13, Okamura described the new species *Archaeoanas japonica*, a duck from the Silurian strata of the Kitagami mountain range. Its resemblance to modern ducks is incontrovertible. His illustration (Figure 1) clearly shows the specimen "in state of cramp through shock by being buried alive during the Silurian Period."[1] The specimen measures just 9.2 mm long, clear evidence for the derivation of modern forms from tiny ancestors.

Okamura compares *Archaeoanas japonica* to the celebrated fossils of the Jurassic proto-bird *Archaeopteryx lithographica*, concluding,

> the hypothetical ancestor of Archaeopteryx must be changed to the true birds of the Archaeonas and others, and the former must belong to the mere Reptilia with membranous wings, Pterosauria.[2]

Thus in a single sentence Okamura shakes from its roost one of the most well-known fossils, providing evidence for the direct lineage of modern birds not from *Archaeopteryx*, not from dinosaurs, but from birds of quite diminutive size that otherwise wholly resemble their modern counterparts.

A discovery as remarkable as that of the Silurian duck, showing a direct biological lineage to the modern avifauna, would be a life's accomplishment for an ordinary scientist. But Okamura's observations and discoveries continue. In *OROFL* 14, he synthesizes, in a paper of 182 pages and more than 1,000 photomicrographs, the evidence for the origin of vertebrates.

The Source of Life

In *OROFL* 14 (Figure 2) Okamura reports that he had

> collected a great deal of black limestone from the Nagaiwa mountain in Higoroichicho, Ofunado city, Iwate prefecture. Polished specimens were examined microscopically and those found were chiefly fossilized land, rather than marine vertebrates of 92 species especially many fossilized human beings and aquatic and land plants, mixed with old continental Paleozoic fossils: . . . all creatures from the Cambrian to the Silurian periods.[3]

He continues, "All these living things had a ministructure of about 1.0–5.0 mm in size and nevertheless each [represents] an identical species of the Recent age . . ."[4] He describes minifishes, minireptiles, miniamphibians, minibirds, minimammals, and miniplants.

Not content to relax at the astounding discovery of minimen and miniwomen, Okamura also discloses his discovery of groups of "protominimen," with bodies similar to those of dragons. He further demonstrates that he has specimens that show "a continuous and systematic metamorphosis of the protominiman."[5]

Okamura sums his brilliant stroke in a single sentence: "The old Nagaiwa mountain was the cradle of all creatures on the Earth."[6]

Lions, and Tigers, and Bears

Most of the forms that Okamura has described are cited as new species and subspecies of modern genera and species; their resemblances are clear. Most of the new fossil taxa are subspecies of modern species;

Original Report of the Okamura Fossil Laboratory

No. XIV

極 東 ミ ニ 生 物 の 時 代

PERIOD OF THE FAR EASTERN MINICREATURES

岡村化石研究所長　　岡 村 長 之 助 著
455　名古屋市港区甚兵衛通5丁目12番地

Written by CHONOSUKE OKAMURA

Director of Okamura Fossil Laboratory

5-12, Jinbeidori, Minato-ku, Nagoya, 455, Japan

此の論文は一般読者の為、特に平易な日本語と英語とを以て述べたものである。

This paper was given as far as possible in non-technical Japanese and non-technical English.

此の論文は次の日本古生物学会で報告されたものである。

1. 第 119 回例会 ・・・・・・・・・・1977年 6 月 18 日・・・・・・・・・・・静 岡 大学
2. 第 120 〃 ・・・・・・・・・・・・ 〃 10月 16 日・・・・・・・・・・・熊 本 大学
3. 1978年年会 ・・・・・・・・・・・1978年 1 月 20 日・・・・・・・・・・・京 都 大学
4. 第 121 回例会 ・・・・・・・・・・ 〃 6 月 3 日・・・・・・・・・・・筑 波 大学
5. 第 122 〃 ・・・・・・・・・・・・ 〃 10月 14 日・・・・・・・・・・・山 形 大学
6. 1979年年会 ・・・・・・・・・・・1979年 1 月 22 日・・・・・・・・・・・福 岡 大学
7. 第 123 回例会 ・・・・・・・・・・ 〃 6 月 9 日・・・・・・・・・・・金 沢 大学
8. 第 124 〃 ・・・・・・・・・・・・ 〃 10月 20 日・・・・・・・・・・・名古屋大学
9. 1980年年会 ・・・・・・・・・・・1980年 1 月 26 日・・・・・・・・・・・筑 波 大学

This paper was reported at the Paleontological Society of Japan.

1. At the 119th regular meeting on June 18, 1977 at Shizuoka University.
2. At the 120th regular meeting on Oct. 16, 1977 at Kumamoto University.
3. At the annual meeting on Jan. 20, 1978 at Kyoto University.
4. At the 121th regular meeting on June 3, 1978 at Tsukuba University.
5. At the 122th regular meeting on Oct. 14, 1978 at Yamagata University.
6. At the annual meeting on Jan. 22, 1979 at Fukuoka University.
7. At the 123th regular meeting on June 9, 1979 at Kanazawa University.
8. At the 124th regular meeting on Oct. 20, 1979 at Nagoya University.
9. At the annual meeting on Jan. 26, 1980 at Tsukuba University.

Figure 2: Title page of OROFL 14.

for example, the minilynx (*Lynx lynx minilorientalis*), the minigorilla (*Gorilla gorilla minilorientalis*), the minicamel (*Camelus dromedarius minilorientalus*), the mini-polar bear (*Thalarctos maritimus minilorientalus*), and the mini-common dog (*Canis familiaris minilorientalis*), whose "features were similar to those of a St. Bernard [sic] dog, but the length was only 0.5 mm."[7]

Okamura also has discovered the ancestral forms of extinct species; for example, a minipteradactyl (*Pteradactylus spectabilis minilorientalus*) and a children's perennial favorite, the minibrontosaurus (*Brontosaurus excelsus minilorientalus*) (Figure 3).

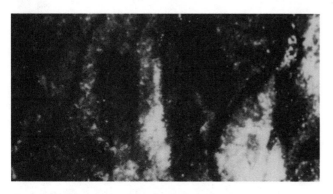

Figure 3: Photomicrograph of a minibrontosaurus, Brontosaurus excelsus minilorientalus (OROFL 14, plate 38, figure g).

In most of the descriptions, Okamura combines the power of scientific deduction with sympathetic observation. For example, in his description of *Lynx lynx minilorientalis* he notes:

> Some look frightened in anger against a sudden convulsion of nature while others are indifferent or even have sunk their heads on their breasts having lost the power of resistance. These are remnant remained forms of psychical movement showing the degree of development of intelligence.[8]

Here we see unique and inspired insight not only into systematic paleontology, but with consideration of the academic conflict of environment vs. heredity in the evolution of intelligence.

New genera for a few kinds of minifossils have been reserved for those that lack modern counterparts. However, in one instance, when faced with a totally new form of reptile, Okamura arrives at a still yet more remarkable discovery, which provides us with some insight into his proclivity for deduction.

The Line Between Sorcery and Research

Okamura left for the field and travelled deep into the mountains of Gifu Prefecture, to learn more about the Yokozuchi, "which had been left long alone as an unknown living thing." He interviewed road workers and elderly residents about this creature, and he was directed to a book, the *History of Tokuyama Village*, "published May 27, 1973 by Mr. T. Neo, Chief of the Tokuyama Village."[9] Okamura quotes from page 77 of Neo's book: "there lives a

rare poisonous snake called Yokozuchi, which cannot be found in any picture book." Neo's book gives a crude line drawing, which Okamura reproduces.[10]

Based on this evidence, Okamura describes and names the reclusive snake *Yokozuchius yokozuchius*, and the Silurian minisnake *Y. y. minilorientalis*.

What is especially remarkable is that Okamura is able to discover a previously undescribed *modern* species by first identifying its approximately 420 million-year-old fossil ancestor. This may open a path of research toward solving the riddle of the Himalayan Yetis, and the Bigfeet of the American Northwest, among other ostensibly legendary creatures.

To this end, Okamura achieves the goal of scientifically describing dragons, naming the group (Draconae) and including in it 18 new genera. He also describes the life habits and physiology of dragons. It is especially noteworthy that dragons are often camouflaged; and the resemblance of the rope-shaped kind of camouflage (Figure 4) to the test of a foraminifera no doubt will call for a revision of studies of these biostratigraphically important fossils. Similarly, paleobotanists and invertebrate paleontologists alike will need to heed Okamura's discovery of the tiny fossil tree, *Lepidodendron minilorientoanulus*, whose resemblance to the internal structure of corals may force revision in these disciplines.

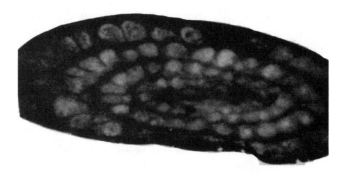

Figure 4: Photomicrograph showing the "rope" style of dragon camouflage (OROFL 14, plate 55, figure 55e).

The Dawn of Man

Certainly, Okamura's greatest claim to fame is the discovery of the miniman, *Homo sapiens minilorientales*.[11] In a lengthy and meticulous anatomical discussion, illustrated with hundreds of photomicrographs (e.g., Fig. 5), the earliest ancestors of humans are described. "The Nagaiwa miniman had a stature

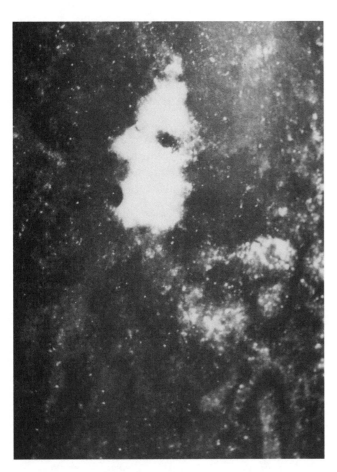

Figure 5: Photomicrograph of a specimen of Homo sapiens minilorientales, a Nagaiwa miniwoman "about 30 years of age . . . [and who] seems to be wearing a mantle of some kind on which many small dragons had been pasted, perhaps an after death phenomenon" (OROFL 14:272, plate 62).

of only 1/350 that of the Recent man but with the same shape."[12] The tools of these minipeople are described, too, including "one of the first metallic implements."[13]

Paleoenvironmental aspects are not overlooked in Okamura's detailed descriptions of the minipeople. Representative samples of his interpretations are:

> . . . the miniman had a size like that of the Recent small ant and probably dwelt in caves after some development, or lived in simple type of houses constructed of plates of calcite or something similar. Furthermore, they knew letters and how to make cement by baking of calcite, and how to make china . . .[14]

All the women in Fig. 70 have closed mouths and [are] seen to be undergoing pain by being buried alive in boiling mud, while the old woman in figure 1 has a wide open mouth looking like one who has lost her senses . . .[15]

They were polytheists and had many idols installed.[16]

There are detected "the oldest hair styles,[17] "a quick-footed Nagaiwa miniwoman [who was] probably a hard worker,"[18] a miniwoman who "seems to have been a person of noble rank,"[19] and the "Head of a miniman in the alimentary canal of a dragon."[20]

The Nagaiwa minipeople were artisans, too, producing a broad variety of sculpture. "What may be regarded as the most elaborate piece of work" is that of a "full-length portrait of a woman sitting on the neck of a dragon," who "may be putting on a hat." Okamura "presumes this to be some kind of goddess," whose "mammae seem to be quite swollen and sagging a little."[21]

Natural Selection in the Early Paleozoic

The Nagaiwa miniworld was not idyllic. Okamura illustrates a "Close nestling protominiman and protominiwoman . . . both defying a dragon,"[22] a "Dragon strangling a girl,"[23] and a "Miniman offering a sacrifice to a brutal dragon,"[24] among other insightful tableaux. However, this review cannot even begin to place in proper perspective the wealth of detail seen by Okamura.

The relationship between minimen and dragons appears to have been not mutually beneficial, if Okamura's interpretations are correct.

> From what the author could determine, the minimen lived in the ancient times having a high intellectual level with only flat nails for protecting themselves. Even if they grasped poles, using their free upper limbs, or used primitive metallic arms which seem to have existed, or hurled simply processed stones, it would have been most difficult to escape from the gluttoneous desire of countless flesh-eating hungry dragons.[25]

The earlier forms of minipeople were without hands, but

> it would have made no difference if there had been a hand to hand fight with dragons, they still would have been defeated without the least resistance. The dragons would have mortally wounded them and crushed their bodies.[26]

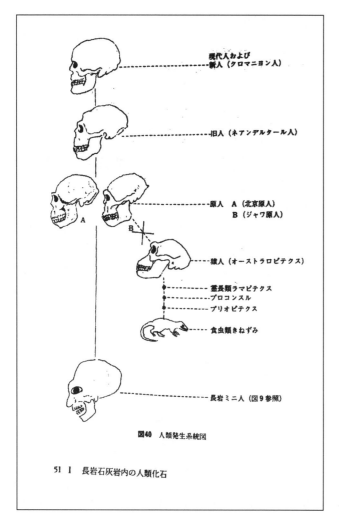

現代人および
新人（クロマニヨン人）

旧人（ネアンデルタール人）

原人　A（北京原人）
　　　B（ジャワ原人）

猿人（オーストラロピテクス）

霊長類ラマピテクス
プロコンスル
プリオピテクス

食虫類きねずみ

長岩ミニ人（図9 参照）

図40　人類発生系統図

51　I　長岩石灰岩内の人類化石

Figure 6: Phylogeny of human ancestry, illustrating the direct link to early Paleozoic minimen, and not from tetrapods as previously believed. (Okamura 1983?, plate 51)

Yet Okamura's startling observations were not totally detatched from emotion; he wrote, "The author will do his best to comfort their departed spirits."[27]

Implications for Evolution

The author has discovered more than 93 species of the Nagaiwa minivertebratae, all of which had the same corresponding species in the extinct or Recent period and there were no new species, that is, there was no big evolution, nor metamorphosis over very long times in vertebratae bodies[28] [Figure 6].

There have been no changes in the bodies of mankind since the Silurian period . . . except for a growth in stature from 3.5 mm to 1,700 mm.[29]

Postscript

The Okamura Fossil Laboratory has apparently not produced further work since ca. 1987. Chonosuke Okamura himself appears to have retired to a secluded life. His carefully detailed work simply drifted into obscurity, a sad example of inadequate publicity. It is clear that revisionist forces in paleontology and anthropology must take proper account of the overlooked, extremely well-documented evidence of Okamura's minivertebrates. The scientific community should place this man's work in the light that it deserves.

Selected Publications

1. *"Archaeoanas japonica,"* Chonosuke Okamura, *Report of the Okamura Fossil Laboratory*, Nagoya, Okamura Fossil Laboratory, no. 13, 1977, pp. 157–63, plate 31.
2. "Period of the Far Eastern minicreatures," Chonosuke Okamura, *Report of the Okamura Fossil Laboratory*, no. 14, Okamura Fossil Laboratory, Nagoya, 1980, pp. 165–346, plates 32–107.
3. Chonosuke Okamura, Okamura Fossil Laboratory, Nagoya, 1983? [Book entirely in Japanese; with errata sheet for *OROFL* 14].
4. *New Facts: Homo and All Vertebrata Were Born Simultaneously in the Former Paleozoic in Japan*, Chonosuke Okamura, Okamura Fossil Laboratory, Nagoya, 1987.

Notes

1. *OROFL* 13, p. 51
2. p. 50.
3. *OROFL* 14, p. 174
4. p. 174
5. p. 305
6. p. 339
7. p. 262
8. p. 257
9. p. 226
10. p. 226
11. p. 269
12. p. 271
13. p. 297
14. p. 271
15. p. 289
16. p. 304
17. p. 279
18. fig. 64b
19. fig. 63f

20. fig. 65d
21. p. 301
22. fig. 103
23. fig. 65h
24. fig. 66
25. p. 273
26. p. 273
27. p. 276
28. p. 344
29. p. 272

Appendix

The following institutions hold copies of the *Reports of the Okamura Fossil Laboratory*:

Academy of Natural Sciences (Philadelphia, PA)
Colorado School of Mines
Cornell University
Denver Public Library
Field Museum of Natural History
Harvard University, Museum of Comparative Zoology
Kent State University
Pell Marine Science Library (Narragansett, RI)
Smithsonian Institution
U.S. Geological Survey (Reston, VA)
University of California at Los Angeles
University of California at San Diego
University of Houston
University of Texas at Austin
University of Wyoming

CHAPTER 4

Astronomy, Physics and Food

Scientists are obsessed, ultimately, with two things: the universe and food. Some researchers have managed to combine their twin passions, as exemplified here by articles such as "The Effects of Peanut Butter on the Rotation of the Earth," "The Aerodynamics of Potato Chips" (which was written by, among others, scientists from NASA), "Nanotechnology and the Physical Limits of Toastability," and "The Laser Cheese Raclette." The last-named demonstrates how expensive scientific equipment can be used to prepare a refined and delicious snack.

The universe is a big place. Astronomers are always trying to figure out just how big it is and how we fit into it. Philosopher George Engelbretsen's essay "Mondocentrism," will give comfort to those who yearn for a simple answer.

Physics is considered to be the queen of the sciences. It is also considered to be the father of all the sciences. Some have called it the foundation of all the sciences. Scientists, especially physicists, like to give things labels, nice impressive labels. Steve Nadis's masterful search for the one, true meaning of the phrase "In Search of the Holy Grail" brings this sophomoric habit to its, yes, logical conclusion.

For those who require UFOs and space aliens, we present both research and useful information. Leonard X. Finegold has done a straightforward analysis of your risk of being abducted by aliens. Unlike most of what appears in the popular press, this is genuine, straightforward scientific analysis. And our chart of scheduled UFO sightings is much more useful than the insubstantial after-the-fact sighting "reports" that stir so much fuss.

Chaos? Everyone writes about it. So do we. Inaudi, Colonna de Lega, Di Tullio, Forno, Jacquot, Lehmann, Monti, and Vurpillot's report, "Chaos: Evidence for the Butterfly Effect," is the first conclusive evidence that the famous "butterfly effect" is not just a delicate, pretty metaphor.

Interspersed throughout the chapter, you will find helpings from several of *AIR*'s regular features. "May We Recommend" summarizes genuine research reports that our readers have spotted and sent to us—these citations are perhaps the finest and noblest things to appear in *AIR*, and you will find more of them scattered throughout this book. "Sleep Research Update" speaks for itself. "Scientific Gossip" is just that—gossip that can be digested or passed at your own discretion. "*AIR* Vents" is a collection of what other publications call "Letters to the Editor." We generally publish a lot of them; in this book we've included a tantalizing taste.

Chaos: Evidence for the Butterfly Effect

by D. Inaudi,* X. Colonna de Lega,*
A. Di Tullio,* C. Forno,† P. Jacquot,*
M. Lehmann,* Max Monti,* S. Vurpillot*

* Laboratory of stress analysis, Swiss Federal Institute of Technology, Lausanne, Switzerland
† Invited Professor from City University, London, England

This appeared in *AIR* 1:6 (November/December 1995).

Note: The individual butterfly responsible for rainfalls in Paris (France) was found in Lausanne (Switzerland). This paper presents the experimental results leading to this finding, as well as an ethical and philosophical discussion on the issues raised by this discovery.

Background

Some phenomena are so complex that even miniscule actions can have enormous unpredictable consequences. From chaos theory,[1] it is well known that a single butterfly[2] wingbeat can produce catastrophic reactions in distant countries. Different formulations of this natural law refer to the formation of tornadoes in the USA, thunderstorms in Japan or rainfalls in Paris.[3] The question whether similar meteorological events in the same region are the result of the agitation of a single individual insect has not, to the authors knowledge, ever been raised.

Figure 1: The butterfly responsible for all rainfalls in Paris.

Methodology

This first feasibility test focused on rainfalls in Paris. This city was chosen because reliable meteorological data were available for it.

Ten butterflies were chosen to represent, in politically correct proportions, the species distribution of these insects in Switzerland. To avoid any possible contamination of the results, the tests were conducted using a double-blind procedure. The butterflies were not told they were taking part in a scientific experiment.

Each morning, one of the authors phoned (at the lab's expense) his girlfriend in Paris to inquire about the weather. To ensure a reliable result the observations extended daily over at least one hour during which permanent telephonic contact was kept. At the same time, a second randomly chosen researcher, unaware of the contents of the phone call, observed the activities of the ten butterflies.

After all the relevant information was gathered, the two outcomes were compared.

Whenever the sun was shining in Paris the entomological data was discarded since the butterfly responsible for rain could be flying for other reasons on sunny days. In the case of wet weather the activity or inactivity of each specimen was recorded for further analysis.

The study continued until the first telephone bill reached our Office of Financial Services. It was possible to obtain results over a span of 54 days, 16 of which were rainy.

Experimental Results

Table 1 summarizes the observations of butterfly activities in Lausanne during 16 rainy days in Paris. These results show a clear correlation between the activity of butterfly J. L. (see Figure 1) and rainfalls in Paris. Butterflies Curt and Mr. X showed promising activity patterns but failed to produce rainfalls in a reliable and uninterrupted way. The specimen Max showed no detectable rain-related form of activity.

Since the probability of a butterfly being active at the right time for all 16 days is only one in $2^{16} = 65,536$, we must conclude that there is, indeed, an insect in Lausanne that is producing all rainfalls in Paris.

However, it was not possible, with our limited resources, to ascertain whether the other butterflies in the study were responsible for the weather in other world regions. The authors invite all readers to check the activity patterns of the other butterflies against the weather records in their own countries and inform us about any correlations.

Table 1

	Al	Curt	C.H.	Dan	J.L.	Mat	Max	Ray	Sam	Mr. X
4/Feb/95	X	X			X	X			X	X
4/Feb/95	X	X	X	X	X					X
8/Feb/95		X			X	X				X
18/Feb/95	X	X			X				X	
19/Feb/95	X	X			X	X			X	X
20/Feb/95	X	X			X				X	X
24/Feb/95		X	X	X	X	X			X	X
29/Feb/95		X		X	X				X	X
1/Mar/95		X	X	X	X	X				
7/Mar/95		2		X	X				X	X
8/Mar/95	X	3	X	X	X	X		X		X
13/Mar/95	X				X			X	X	X
14/Mar/95	X			X	X	X		X		X
21/Mar/95	X				X			X	X	X
22/Mar/95	X			X	X	X		X	X	X
25/Mar/95			X	X	X			X	X	
Hit/Miss	10/6	9/7	5/11	9/7	16/0	8/8	0/16	7/9	10/6	13/3

Table 1: Activity of the individual butterflies on each of the sixteen days on which it rained in Paris. The insects were identified by pseudonyms (to protect privacy). An "X" indicates some form of activity (e.g., wingbeats). 2 indicates probably dead. 3 indicates certainly dead.

Discussion

The meteorological implications of this discovery are, naturally, of great interest, but its ethical, sociological, and commercial consequences must all be considered.

Certain questions come to the fore. First, would it be possible to control the weather in Paris by restraining the butterfly movements (by invoking either nail or pin theory)? If so, then the owner of this insect could gain great influence in political circles (e.g. some analysts argue that left-party supporters are more likely to cast their votes on a rainy day). The commercial potential of such a discovery would be large. It would be possible to sell a sunny day for special occasions such as a high-class wedding, the Roland Garros tennis final, or Mr. Du Pont's car wash day.

There are also ethical and philosophical questions. Must the J. L. butterfly be considered harmful and thus be eliminated, or is it an essential part of the Paris ecosystem? What would happen if its life were brought to a sudden and violent end? The authors could not reach agreement on this point. Some argued that if a simple wingbeat creates a strong rainfall in Paris, the violent death of the butterfly could produce a cataclysm of great proportions. Other think that the soul of the insect would simply migrate to another individual. This later theory is suggested by the experimental data: one can see an apparent transfer of activity to butterfly Ray after the death of butterfly Curt. Ray could therefore be the reincarnation of Curt.

Mindful of the potential havoc our work could invite, we have formed an ethical committee to protect the meteorologically active insects from any form of manipulation by individuals and political, military, religious, or sporting organizations.

Conclusions

There is strong experimental evidence of the so-called butterfly effect. The butterfly responsible for rainfalls in Paris, France was found in Lausanne, Switzerland. This finding has an error margin below 0.1%. The consequences of this discovery are worthy of further study.

The authors are indebted to Mother Nature for giving them the opportunity to study such interesting phenomena. This research project was involuntarily sponsored by the Swiss Telecom. The authors would like to express their deep regret to the inhabitants of Paris for the heavy rainfalls produced on the 10th of May, when the photograph (here labeled Figure 1) was shot.

References

1. *Chaos Theory Journal*, all volumes, all issues, all pages.
2. "Discovery of a new flying species," *Journal of Applied Entomology*, 3rd stone, 1238 B.C.
3. *Time Atlas of the World*, 1995 edition.

The Ubiquitous Holy Grail

by Steve Nadis
Cambridge, Massachusetts

This appeared in *AIR* 2:2 (March/April 1996).

The term "holy grail" is almost ubiquitous in the scientific literature (e.g., the "holy grail of hair replacement therapy" or the "holy grail of high-energy metaphysics"). Impressed by the apparent importance of this phrase, I set out to find all references to the term in the contemporary periodical literature and then to deduce its meaning from "contextual" and/or "other" clues.

The First Step

As a first step, I went to the public library and checked the card catalog. Unfortunately, this "catalog" no longer existed. "It's all been computerized," explained the librarian, June, whom I vaguely recalled from my book-reading days. She directed me to a machine called "InfoTrac." After describing the capabilities of this electronic fact-monger, June typed the words "holy grail." The machine was quiet for a moment, then flashed on the screen: "737 hits!" It asked, "Do you want to narrow the search?" I replied, on the contrary, by all means expand it. InfoTrac, however, could not come up with more than the aforementioned 737 "hits," as previously mentioned. I then had a printout made, a necessary, albeit slow, process that set me back 738 quarters—737 for the copies, plus one for a misprint.

Cable, Cardiology, Car Grails

The 737 references found in the "contemporary" periodical literature are as interesting as they are instructive (see Table 1). Rather than reproduce

Clues in the search for the Holy Grail. Documontage: Stephen Drew.

Table 1 in its entirety, I shall instead mention a few highlights. There was, for instance, the "Holy Grail of cable television" which evidently means "video on demand, the ability to order a movie and play it as though [it] was in your VCR, pausing, rewinding, and fast-forwarding . . ."[1] The "long-sought holy grail of materials scientists," on the other hand, consists of a new synthetic material (not-yet-synthesized) that is "actually harder than diamond."[2] A cardiologist at the Harvard Medical School offered a hard-and-fast definition that was considerably softer than diamond: "The holy grail is for a single test to do everything I need to know."[2a] No less a journal than *Scientific American* averred that the holy grail is, on the contrary, a battery that could "safely power a car for 300 miles on one charge."[3] This contention conflicted with a previous statement in the very same journal arguing that the elusive grail is none other than the Higgs boson, a particle whose precise nature is best left shrouded in a veil of ignominy.[4,4a]

Waves, Hot, Cold, and a Heavenly Grail

Physics, of course, is a realm of almost limitless grails. The "H.G." moniker has been variously ascribed to the "theory of everything" (a.k.a. TOE),[5] the search for gravity waves,[6] the creation of a "Bose-Einstein condensate,"[6a] the decay of the proton into charged particles,[6b] and the achievement of a self-sustaining nuclear fusion reaction, hot or cold, above or below the tabletop[7,8,9]—an accomplishment that has also been claimed by thermodynamicists as "Energy's Holy Grail."[10] There are some subtle distinctions of which the uninitiated may not be aware. The top quark, for example, is not the treasured H.G., but rather the "Great White Whale of Physics."[11] And the "ultimate theory of everything" (UTOE), which supersedes the mere "theory of everything," is actually the "Golden Fleece," not the Holy G.[12]

Cosmology is another area replete with grails for every occasion. To some practitioners, the determination of the Hubble constant (and, by extension, the age of the universe) is the field's H.G.[13] Others have applied the term to the discovery of "primordial wrinkles" in the fabric of space-time, while others still call the latter "the handwriting of God" and/or the "missing link," preferring to reserve the holy grail appellation for loftier concerns.[14,15,16] One should note, again, that the "fingerprint of God" and

the "handwriting of God" bear no relation whatsoever to the "fingers of God."[17]

Smart Grails, Long-Lived Grails

Special mention should also be paid to artificial intelligence[18a] and "immortality," which was called the "holy grail of longevity" on no fewer than seven occasions (a number we shall return to shortly).[18b]

So what do we make of this "chimerical entity," the holy grail—term used in a dizzying concatenation of circumstances and contexts? After a systematic review of the data, I reached several broad conclusions. First, it is virtually impossible to infer a single, "hard-and-fast" meaning. "Holy grail," it would appear, has the peculiar capacity to mean many things to many people, if not all things to all people. This chameleon-like characteristic makes for a very slippery quarry indeed.

Grail in a Garage

Confusion on this matter was compounded, rather than alleviated, by recent reports in the popular press that the mythical grail had been found "once and for all." One British eccentric claimed to have stumbled upon a platter in his cousin Ginger's attic which was "unquestionably the grail in question." Another putative archaeologist found a mug at a rummage sale—a rugby trophy, in point of fact—that bore a "distinct likeness" to the grail of grails.[18c]

A Blockbuster Clue

It all made for a baffling conundrum. If I were to have any chance of solving this riddle, I would, at the very least, need a new perspective—a new avenue of attack. As the public library was of no further use to me, I moved on to another scholarly venue—the local outlet of Blockbuster Video. In the Blockbuster database, I found reference to a movie, "Monty Python and the Holy Grail," that, perhaps, held the key to this age-old mystery. Unfortunately, the movie was already checked out.

Nevertheless, I did glean some important information from the Blockbuster salesperson, who struck me as a resourceful, if not particularly knowledgeable, young man. He had not seen the movie for "quite some time, mind you,"[19] yet seemed to recall that it had something to do with the search for "a religious artifact of some sort." That, to first approximation, may be as close as we can arrive, at present, to the meaning of the elusive grail—a "religious artifact of some sort."

Something Completely Different

I was about to conclude my investigation at this juncture, when I had a chance encounter with a Blockbuster patron—a young woman who waited patiently through the entire exchange in the hopes of checking out the Nazi dance film, "Swing Kids." This informant (call her "Lady X" for the sake of confidentiality) passed on this information to me, although I cannot vouch for its veracity. The "snake movie" (python?), she submitted, starred a certain John Cleese, whose name rhymes with "keys." That struck me as potentially significant. More to the point, she said, this same "J. C." was born on the 7th day of the 7th month in such and such a year "A.D." (the exact date being of no special import.) A pattern was emerging, a striking juxtaposition of "7's," one almost immediately after the next. That "7," coincidentally, happens to be the number of days in a typical week (leap-year notwithstanding). Furthermore, that also happens to be the same number of days, give or take, that our good Lord took to create the Earth and everything on it, including the mysterious grail. Future investigators would do well to follow up on this connection.[20]

Notes

1. "For the Couch Potato . . . ," George Judson, *New York Times*, August 20, 1995.
2. "Nearly diamond-hard substance is synthesized," David Chandler, *Boston Globe*, February 25, 1995.
2a. "Smaller firms developing new types of medical light," *Boston Globe*, October 19, 1994.
3. "Bettering Batteries," Sasha Nemecek, *Scientific American*, November 1994, p. 106.
4. "Lone Star Science," John Horgan, *Scientific American*, January 1989, p. 17.
4a. "Particle accelerators and matters of faith," Chet Raymo, *Boston Globe*, January 25, 1993, p. 26.
5. "Taking a Quantum Leap," D. Smith, *Bostonia*, July/August 1988.
6. "Gravity Wave Detectors," Theresa Hitchens, *Smithsonian News Service*, July 1988.
6a. "Physics 'Holy Grail' Finally Captured," C. Wu, *Science News*, July 15, 1995.
6b. "'Super' Japanese Site Gears Up to Solve Neutrino Puzzle," Dennis Normile, *Science*, November 3, 1995, p. 729.
7. "Cold Fusion," Jerry Bishop, *Popular Science*, August 1993, p. 47.
8. "Cold fusion . . . ,"D. Chandler, *Boston Globe*, December 11, 1989, p. 46.
9. "Physicists Discuss Fusion Breakthrough," Elizabeth Thomson, *Tech Talk*, January 5, 1994.
10. "Energy's Holy Grail," Robin Johnson, *Research Horizons*, Winter 1990, p. 9.
11. "A unit of matter may be found," Douglas Birch, *Boston Globe*, December 29, 1992, p. 6.
12. "Search Quickens for Ultimate Particles," Malcolm W. Brown, *New York Times*, July 19, 1988, p. C13.
13. "Search for Cosmology's Holy Grail," Ron Cowen, *Science News*, October 8, 1994.
14. "Scientists Report Profound Insight. . . ." John Noble Wilford, *New York Times*, April 24, 1992.
15. "The Handwriting of God," S. Begley, *Newsweek*, May 4, 1992, p. 76.
16. "COBE Causes Big Bang in Cosmology," M. Stroh, *Science News*, May 2, 1992.
17. "A New Map of the Universe," A. Dyer, *Astronomy*, April 1993, p. 44.
18a. "School for Robots," Peter J. Howe, *Boston Globe*, October 20, 1995, pp. 29–36.
18b. The sedulous investigator will have no problem digging up these citations.
18c. *ibid.*
19. An indefinite period generally exceeding a year.
20. I can and will swear on the holy grail to the accuracy of everything contained herein. This type of personal guarantee, I believe, makes further citations not only superfluous, but holy unnecessary.

Sleep Research Update
by Yuska-Marie Paskevitch

RESEARCH GROUP 1

KD is sleeping with RM.
RM is sleeping with PI.
PI is sleeping with RK.
RK is sleeping with WB.
WB is sleeping with GG.
GG is sleeping with FP.
FP is sleeping with KD.

RESEARCH GROUP 3

TFD reports a string of disappointing results. She is seeking an improved research protocol.

RESEARCH GROUP 4

DS is breaking in a new graduate student.

RESEARCH GROUP 7

FL has been experimenting with hair dyes.

RESEARCH GROUP 7

KD lost his research grant, and is not sleeping with anyone.

These results are collected from various issues of *AIR*.

Scheduled UFO Sightings

March 15	Matmaya, Tunisia
April 24	Wieliczka, Poland
April 25	Beppu, Japan
June 19	Lambert Glacier, Antarctica
July 1	Whitehorse, Yukon Territory, Canada
August 2	Puno, Peru
October 5	Simi Valley, California, USA
December 24	Moscow, Russia

Please consult local authorities for exact times.

This appeared in *AIR* 2:2 (March/April 1996).

A Curious Particle Accelerator in Switzerland

This photograph graced the cover of *AIR* 1:3 (May/June 1995).

This photograph was found in a storage room at the CERN high energy physics research facility near Geneva, Switzerland. The number "1952" was written on the back. The apparatus is thought to be an obsolete type of particle accelerator. Photo: Robert Richard Smith.

The Laser Cheese Raclette

by A. Zryd, T. Liechti, and J. D. Wagniere

Laser Center for Material Treatment
Swiss Federal Institute of Technology
Dept. of Materials Engineering
CH-1015 Lausanne, Switzerland

This appeared in *AIR* 1:3 (May/June 1995).

"Raclette" is an archaic dish from the Swiss Alps, consisting of molten hard cheese. It differs from the well-known "Swiss fondue" by its preparation: only the upper layer of the whole piece of cheese is molten by exposition to a heat source, the insides are scraped out onto a plate, and eaten with potatoes and dry white wine. The process can be repeated as long as there remains cold cheese. Typical heat sources include wood combustion[1] and, more recently, heat produced from electricity via the Joule effect.[2] The first of these techniques presents the inconvenience of not always being available in a modern environment. Both traditional methods require hand injection of pepper, which slows down the whole process. This paper presents a novel technique that eliminates both disadvantages, using equipment that is common in most modern laboratories.

Experimental Set-up

A commercially available Swiss cheese was chosen as a test sample and ground black pepper was used for injection. A 1.5 kilowatt continuous wave CO_2 laser was defocused on the cheese to melt the upper layer. The piece of cheese was displaced under the stationary beam using a numerically controlled X-Y table. A picture of the experimental setup is given in Figure 1. The processing conditions are presented in Table 1.

The pepper was deposited using a blown-powder technique, described elsewhere.[3] For this purpose a Plasma-Tech Twin 10C powder supply was used, with Argon as a carrier gas. The influence of the particle size distribution of the pepper was stud-

Figure 1: Laser cheese remelting facilities. The cheese can be seen in the lower part of the picture (half-moon shape). The laser beam comes from above, through the optical system. The small tube coming out of the picture is the cross jet protecting the lenses from pepper particles. Behind is the vent exhaust to remove the smoke and smell.

Table 1

Processing parameters of cheese remelting by CO₂ laser.

Laser power	250 W
Beam diameter	5 cm
Scanning speed	4500–6000 mm/min

Table 1: Processing parameters of cheese remelting by CO₂ laser.

ied. It has been found that, after serving, pepper with a grain size lower than 30 micrometers has a negative influence on the powder fluidity due to electrostatic charging. The best results in terms of transportation and taste were found for a particle size between 40 and 100 micrometers (Figure 2).

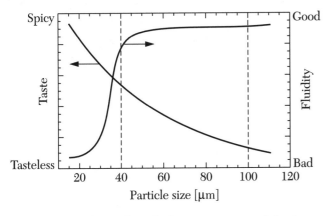

Figure 2: Analysis of crushed pepper: taste and fluidity vs. particle size.

Experimental Results and Discussion

Efficiency

Due to the circular shape of the beam and the limitation in focal size, the whole rectangular cheese surface could not be irradiated in one path. Hence scanning was necessary and had to be optimized with respect to the near top hat mode of the beam.

For a good raclette, the cheese has to heat up to 100°C (certainly not more than 180°C, above which it calcinates) in a thickness of 1–3 mm. As the entire incident power is absorbed in a layer a few microns

thick, and as cheese has a low thermal conductivity, the energy density (power density multiplied by the interaction time) is limited. Therefore the power density has to be reduced to values typical for conventional electrical furnaces.

For this reason the highest laser power available cannot be used. Heating cheese by CO₂ laser (with a wavelength of 10.6 microns) produces results similar to those obtained by heating via the Joule effect (which typically involves mostly infra-red emission). However to produce 1.5 kW laser power, an input of 15 kW is necessary. This strongly decreases the energy efficiency of the laser raclette process.

Processing maps

The heating and melting processes have been simulated by other researchers, using a finite-difference model published elsewhere.[4] The results are presented in Figure 3.

The calculated shape coincides remarkably well with our experimental results. We therefore calculated a processing map which allows one to easily determine the essential laser parameters. These include, most prominently, the number of persons to be served, and the kind of cheese. An example of such a map is shown in Figure 4.

Feasibility and quality control

With the above-mentioned processing conditions, raclette can be served satisfactorily to 23 test persons, though at a somewhat slow rate. With this new method, it is easy to obtain the so-called "religieuse"

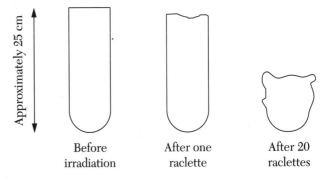

Figure 3: Finite-difference simulation of the shape of the cheese sample during irradiation. As can be seen, the cheese sample experiences a strong deformation due to high temperature creeping.

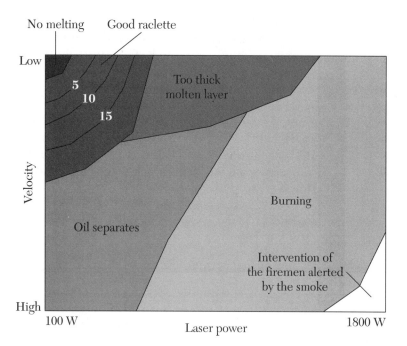

Figure 4: Processing map for conventional raclette Swiss cheese, showing the optimal window for raclette cooking, with the equi-guests lines indicating the best parameters for a given number of guests. A similar map can be established for different kinds of cheese and different optical lengths of the laser.

—the grilled part of the cheese crust, which specialists consider to be the best part of raclette but which is also the most difficult to prepare.

A blind test was conducted to compare the quality of a laser raclette and a conventional raclette. The results of the test are presented in Figure 5. They favor the laser technique. Part of the quality

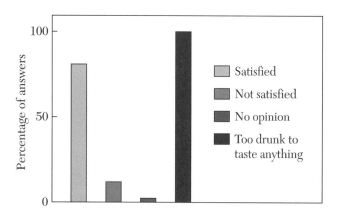

Figure 5: Results of the blind test conducted to check the taste and quality of the laser raclette.

difference is due to the automated dispersion of pepper particles, which produces a much more regular layer, and therefore a more constant taste, than can be obtained by the other methods.

One must also note that this technique, like that of Joule heating, has an advantage over wood heating: it avoids the problem of smoke which has been proven to be carcinogenic.[5]

Economic analysis

Economically, the cost of the laser technique compares favorably to that of more traditional processes. One hour of laser work is assumed to cost 300 Swiss Francs (US $230), including equipment, energy consumption and technician. As processing time is about 120 seconds, the cost of one raclette is estimated to be 10 SFr (US $8). This compares to a price of .08 SFr to prepare one conventional raclette with Joule heating (120 seconds at 15 SFr/hour including equipment energy and unqualified worker) and 0.025 SFr for wood heating. In the latter case, the wood energy is considered to be at zero cost, as usually wood is gathered freely in the forest. This may not be the case in some urban areas.

It must be said that the above calculations do not include secondary costs and environmental aspects, such as air pollution and losses during electricity transportation.

Conclusion

This novel approach to preparing the raclette has given encouraging and tasty results. This is an interesting practical application of the laser. It may open new markets for the Swiss cheese and laser industries. More important, it will provide new opportunities to train laser or surface treatment specialists to perform under adverse conditions involving ethyl vapors.

Acknowledgments

Thanks are due to all the participants in the blind test, in particular Dr. Lang for his active interest in this study. This study has been developed as part of the Development Program for the Alpine region. Their financial support is gratefully acknowledged.

Bibliography

1. "Fire cooking in the underdeveloped mountain regions of the world," E. Whymper, *Journal of the Royal Mountaineering Society*, London, 1853, pp. 2–78.
2. "High temperature emissivity of materials for cheese melting," J. D. Wagnihre, S. Bourba, P. Gilgien, M. Gremaud, O. Hunziker, T. Liechti, L. Poiri, M. Vandyoussefi, A. Zryd, *Acta Formatica*, 1985, vol. 3, pp. 548–57.
3. *High Power Lasers*, R. Dekumbis, H. Mayer, P. Fernandez, A. Niku Lari, ed., Persimmon Press, Oxford, 1989, pp. 289–96.
4. Hoadley A., Rappaz M., *Metallurgical Transactions B*, vol. 23 B, 1962, pp. 631–42.
5. "A statistical study of cyrrhosis and cancer among alpine population," J. Frankenstein, *International Journal of Mountain Medicine*, vol. 234, 1994, pp. 1–54.

Yvette Bassa, the inventor of bright blue Jell-O, accepts the Chemistry Prize in 1992. Bassa's employer, Kraft General Foods, flew Bassa and 20 of her colleagues to the ceremony in the corporate jet. Several other members of Team Jell-O are visible at right, standing behind Nobel Laureate Sheldon Glashow (wearing the white cap) and the ceremony's presiding monarchs, the Swedish Meatball Queen and King. Photo: Luke D'Ancona.

Ig Nobelliana
Words for the ages

"I feel humbled at being singled out for this honor. My achievement is simply the capstone of an immense body of scientific research performed over the past hundred years."

—*Yvette Bassa, winner of the 1992 Ig Nobel Chemistry Prize for inventing bright blue Jell-O.*

Nanotechnology and the Physical Limits of Toastability

by Jim Cser
Applied Breakfast Laboratory
Hillsboro, Oregon

This appeared in *AIR* 1:3 (May/June 1995).

Pretentious Introductory Hype

Some technologies (e.g., Salad Shooters, eight-track tapes, Yugos) have made only a minimal contribution to society, whereas others (e.g., electric guitars, Post-it Notes, Jello 1-2-3) have had far greater impact. Now and then a useful technology catches on, transforming global civilization before our very eyes (see table below).

Right now we are at just such a threshold: namely, the dawn of the Nanotechnology Revolution. Nanotechnology involves building machines that are really, really tiny. As in any true revolution, no one has the slightest idea what is going on. Although various types of nanogadgets have been proposed, these were mostly designed to impress people at parties. Even so, they had only modest amounts of success.

Technological Revolution	Contribution to Civilization
Neolithic Revolution	Medium-sized rocks
Agricultural Revolution	Vegetables
Industrial Revolution	Industry
Microprocessor Revolution	Nintendo
Biotechnology Revolution	Bigger vegetables
and soon...	
Nanotechnology Revolution	Nanotoasters

In any case (especially in this case) talk is cheap. It is time for nanotechnology to prove itself. Clearly, nothing would be more useful than new, improved kitchen appliances. To this end, the author has embarked upon the ultimate quest: an endeavor to stretch the limits of mind, spirit, and my department's funding. I chose to build the world's smallest toaster.

Pseudo-Scientific Hand Waving

What advantages will nanotoasters have over conventional macroscopic toaster technology? First, the savings in counter space will be substantial. Second, since heat transfer scales up and down in tandem with toast area, so does the total heat flux per unit volume of bread scale in inverse proportion to bread dimension. Thus, smaller toast means more efficient toasting. Finally, since nanotoasters will have dimensions smaller than the average wavelength of visible light, there will be no danger of toaster manufacture in awful colors, such as avocado.

Before we can build the world's smallest toaster, however, we must first agree on what a toaster does. The simple answer is that a toaster makes toast. More precisely, a toaster applies heat to a square, flat piece of bread (aspect ratio roughly 10:10:1) until the bread is brown and crunchy. Being able to toast a bagel without having to get it out with a fork would be a big plus, but this is perhaps beyond the

Figure 1: A nanotoaster prepared by the author. Magnification: 56,000. Photo: Stephen Drew.

scope of any conceivable future technology. Pop-Tarts should be accommodated, but should not be a priority, as Pop-Tarts can be manipulated so as to excel in toaster benchmark tests, thus undermining any measurements of overall toaster performance.

One philosophical point that must not be overlooked is that the creation of the world's smallest toaster implies the existence of the world's smallest slice of bread. The smallest quantity of bread that can be sliced and toasted has yet to be experimentally determined. In the quantum limit we must necessarily encounter fundamental toast particles, which the author will unflinchingly designate here as "croutons." It is hoped that quantum toasters will eliminate "crumbs," the discarded byproducts of toast, which cause so many problems at macroscopic scales.

Questionable Experimental Methods

Not surprisingly, the tools needed to fabricate and test nanoappliances are nearly as speculative as the nanoappliances themselves. Fortunately, for a short while (and when no one else was looking), the author had access to an experimental, top-of-the-line Virtual Tachyon Stream Nanoplasty (VTSN) system.[1] According to the manufacturer, VTSN can manipulate undetectably small quantities of matter, using the physical principle of "trust us."

The first test of the VTSN system—making a standard macroscopic piece of toast—was carried out easily. I suspended a slice of bread over the power supplies in the back of the system. The next step was to load the VTSN system with a few grams of paper clips, push the appropriate buttons, and hope for the best.

After ten minutes of loud grinding noises, the system screeched to a halt and released a small puff of white smoke, indicating successful nanotoaster fabrication.

Conveniently, the nanometer-scale slices of bread needed to test the toasters were commercially available through a mail-order scientific bakery supply. A bread size of 50 nm/side was chosen, as the smaller sizes were temporarily out of stock. Both bread and toasters were dumped into a small flask, which was then shaken (not stirred), the theory being that the bread slices would have a natural affinity for their complimentary toaster binding sites.

For the final part of the experiment, the actual nanotoasting process, the reaction flask was placed on a hot plate for a short time. Assuming that the hot plate produced roughly the same amount of heat as a conventional toaster, and using the scaling relation described in the previous section, the toasting time was calculated to be on the order of 100 nanoseconds (several large cups of coffee induced the necessary reflexes for removing the flask). Since there was no evidence of the characteristic "burnt

Figure 2: Three slices of nanotoast. Magnification: 56,000. Photo: Stephen Drew.

toast" odor, the experiment was deemed not to have been an obvious outright failure.

Irrational Conclusions

Because of the nanoscopic size of the toast, the Uncertainty Principle made it impossible to determine exactly the "doneness" of the toast. For that matter, it was also difficult to see whether anything significant had happened at all. However, we must realize that there is a small but finite probability that toasting actually occurred in this equipment.

What does this tell us about the future of nanotechnology? The conventional wisdom currently falls into two camps: either that nanotechnology is the wave of the future, or that nanotechnology is just a big scam. According to my research, the inescapable conclusion is that nanotechnology is both the wave of the future and just a big scam. This fortunate combination of unlimited promise and inherently ambiguous results should generate enough controversy to fuel the engines of science for years to come.

Note

1. The manufacturer, DEI Industries, has since recalled the system for "causality violations" generated, presumably, by a series of bounced checks.

The Aerodynamics of Potato Chips

by Scott Sandford, Jim Ross, Joe Sacco, and Nathaniel Hellerstein

The Aerochip Institute, Mountain View, California

This appeared in *AIR* 1:1 (January/February 1995).

It is a widely held belief that you cannot throw a potato chip. However, it seemed to the authors that the aerodynamics of potato chips must be very sensitive to the Reynolds number.[1] Thus, the truth of the statement that you cannot throw a potato chip must remain in question until actual tests are made at full-scale and at actual flight speeds. The common practice of using wind-tunnel models for the purpose of aerodynamic tests was deemed too difficult and there were the obvious questions of geometric fidelity which always arise in any sort of model construction. In particular, the correct salt distribution, surface roughness, and edge shape would be very difficult to replicate on a model potato chip. For this reason, we decided to subject actual potato chips to a rigorous series of wind-tunnel tests.

Since full-scale testing was determined to be absolutely necessary, the testing was performed in an 80 foot by 120 foot wind tunnel. This facility has two test sections, both powered by a single set of six 40-foot diameter fans (136,000 Hp total). A sketch of the facility is shown in Figure 1. The active test section for a given test is selected by properly positioning vane sets 3, 4, 6, and 7 shown in the figure. This provides for 80′ × 120′ and 40′ × 80′ modes of operation. For this particular test we used a novel, seldom used mode, referred to here as the open warehouse mode, in which the wind-tunnel circuit was configured to blow air through the 40′ × 80′ test section while we used the 80′ × 120′ test section. Since these tests were done during the authors' lunch break, the side model access doors were also left open. The air flow quality in this configuration was excellent and the weather sufficiently good to provide essentially zero turbulence (and zero velocity) in the test section.

Given that the aerodynamic properties of potato chips may depend heavily on their shape, size, weight, etc., we tested a number of different

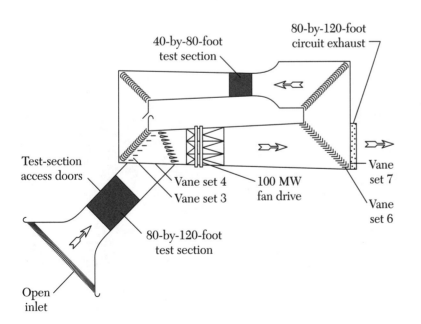

Figure 1: Cross-sectional diagram of the test facility.

Table 1: Chip Types and Average Weights

Chip Type	Fresh Weight (grams)	Stale Weight (grams)
Original Pringles	2.041	2.009
Pringles Sour Cream & Onion	2.092	2.125
Pringles Light Ranch	1.803	1.816
Pringles Corn Chips	2.498	2.628
Pringles Original Ridges	2.726	2.773
El Faro Tortilla Chips	3.130	—
Ruffles Light Choice	1.661	—
Laura Scudders Twin Pack	1.410	—

chip types. (See Table 1 and Figure 2.) In order to separate effects due to shape from those due to weight, we examined a number of different kinds of Pringles chips, as they generally have the same shape while their weight and composition vary. We also compared the results obtained from fresh and stale chips of each type to see if freshness was a factor. The tests were made by launching a series of individual chips of each kind from a height of 13 feet using JSST[2] techniques. Prior to the launching of a new chip type, the Launch Initiator[3] tested the non-stale chips for freshness and his remarks were

recorded.[4] The distance traveled by the chip before impacting the floor of the windtunnel was then measured and recorded. The number of thrown chips of each type was rarely enough to achieve statistically significant results, but was usually determined by the time it took the Launch Initiator to tire of sampling the chip type being tested.

After all the individual chip test had been made, we carried out a further test to see if tight "formation" flight would result in significant decreases in drag. This was found to be the case and was amply demonstrated by the observation that we could throw a bag (or can) of chips much farther than we could throw an individual chip. After the final chip flight test was made, the data were reduced (Figure 3) and thrown into the nearest garbage can. The results of our studies are summarized

Figure 2: The various types of chips tested in the study.

Figure 3: The authors' novel method of data reduction. This photograph was taken inside the wind tunnel at Moffat Field.

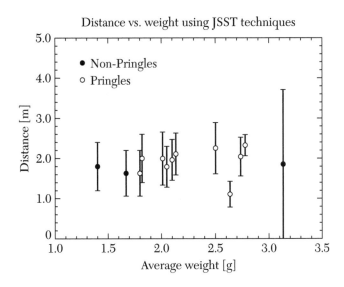

Figure 4: Data indicating that you can throw a potato chip, just not very far.

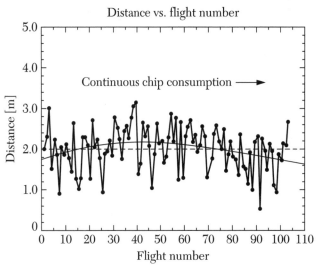

Figure 6: Comprehensive details of the data obtained during this investigation.

in Figures 4, 5, and 6. Figure 4 demonstrates that the adage that "you can't throw a potato chip" is incorrect. Clearly you can throw a potato chip, just not very far. Also, the distance you can throw the

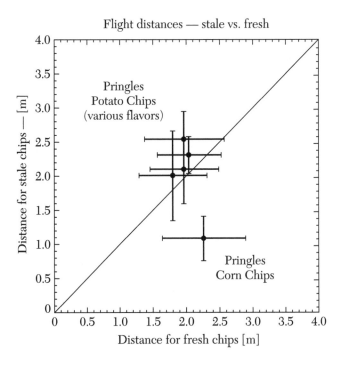

Figure 5: Evidence that stale Pringles potato chips may fly farther than fresh ones. The effect is not statistically significant, however.

chip is largely independent of weight or chip type or shape. Figure 5 indicates that stale Pringles chips may fly farther than fresh Pringles chips, but this conclusion is not statistically significant. Perhaps the stale chips go farther because they have absorbed more water vapor and are heavier.[5] (We note that the opposite is true for Pringles Corn Chips and is statistically significant. We do not understand this and since we are ostensibly testing whether it is possible to throw potato chips [not corn chips], we did not pursue this point.) Finally, Figure 6 shows a plot of flight number vs. distance and covers the entire series of chip types tested. It is included to convince the reader that uniform methods were used throughout the tests. The straight horizontal line is a linear fit to the data and has slope zero, thus demonstrating that steady JSST techniques were used. The curved line is a third order polynomial fit to the data and indicates there was improvement with time on the part of the Launch Initiator, followed by rapid deterioration (also consistent with typical JSST techniques).

Thus, our tests of the adage, "you can't throw a potato chip" yield the following conclusions:

- You can throw a potato chip, just not very far.
- The distance traveled is largely independent of chip weight, shape, or freshness.
- Potato chips travel considerably farther when flying in formation, presumably because such flight decreases overall drag.

- We do not like Pringles Original Ridges Chips. They tasted burnt to us.
- Ruffles Light Choice are light, but still weigh more than the average chip in a Laura Scudders Twin Pack. (See Table 1.)

Manufacturers of potato chips and commuter aircraft should ignore these results at their own peril.

Technical Note

Those who conduct further experiments along this line are advised that, when using a wind tunnel, it is important to let the chips fall where they may.

References

1. Or maybe not. The Reynolds number is a dimensionless coefficient used as an indication of scale for fluid flow; it is fundamental to all viscous fluids. It is also difficult to define in plain language.
2. Joe Sacco Standard Toss. The JSST gave all chips uniform flight potential by providing the unique launch technique best suited for each chip, i.e., no two tosses were alike.
3. Joe Sacco.
4. For example, "Crisp and Tasty," "Mmmmm," "Better than Ruffles Light," etc. (The Launch Initiator refused to taste-verify the stale chips.)
5. An idea in direct opposition to the conclusion from Figure 4.

The Effects of Peanut Butter on the Rotation of the Earth

This appeared in 1993.

Editor's note: With publication of this paper we are hereby amending our longstanding policy regarding co-authors. Previously we rejected any research paper that had more than ten co-authors. Many of our contributors have pointed out that in some fields, especially high energy physics and clinical medical trials, research journals routinely publish papers that have one hundred or more co-authors. Accordingly, we are removing the restriction.

by George August, Ph.D., Anita Balliro, Ph.D., Pier Barnaba, Ph.D., Anne Battis, Ph.D., Constantine Battis, Ph.D., John Battis, Ph.D. Nathaniel Baum, Ph.D., S. Becket, Ph.D., A. G. Bell, Ph.D., Moe Berg, Ph.D., B. J. Bialowski, Ph.D., Edward Biester, Ph.D., Joseph Blair, Ph.D., Ceevah Blatman, Ph.D., Ken Bloom, Ph.D., I. V. Boesky, Ph.D., Dorothy Bondelevitch, Ph.D., Calliope Boratgis, Ph.D., K. T. Boundary, Ph.D., Gerald Brennan, Ph.D., Nuala Broderick, Ph.D., James Burke, Ph.D., Richard Butkus, Ph.D., James Carter, Ph.D., Alexander Cartwright, Ph.D., Caren Cayer, Ph.D., Mary Chung, Ph.D., W. Spencer Churchill, Ph.D., M. Louise Ciccone, Ph.D., Theodore B. Cleaver, Ph.D., Selma Frances Coltin, Ph.D., Carlos Cordeiro, Ph.D., Theodore Crabtree, Ph.D., Samuel Cunningham, Ph.D., James Michael Curley, Ph.D., Gwen Davis, Ph.D., Paul Delamere, Ph.D., R. C. De Bodo, Ph.D., P. deMan, Ph.D., Arthur Derfall, Ph.D., Helen Diver, Ph.D., Edward Doctoroff, Ph.D., Robert Dorson, Ph.D., Wayne Drooks, Ph.D., William Claude Dukinfield, Ph.D., James Durante, Ph.D., Alan Dyson, Ph.D., Raeline Eaton, Ph.D., D. D. Eisenhauer, Ph.D., Kent Fielden, Ph.D., Elizabeth Finch, Ph.D., Raymond Flynn, Ph.D., Charles Follett, Ph.D., Kevin Forshay, Ph.D., George Frazier, Ph.D., Katherine Fulton, Ph.D., R. J. Gambale, Ph.D., Jerome Garcia, Ph.D., Judith Garland, Ph.D., Hannah Gilligan, Ph.D., Daniel Goldfarb, Ph.D., Michael Goldfarb, Ph.D., Archie Goodwin, Ph.D., Yulia Govorushko, Ph.D., Sharon Ph. D. Greene, Ph.D., David W. Griffith, Ph.D., Sheldon Gulbenkian, Ph.D., Frances Gumm, Ph.D., R. O. Guthrie, Ph.D., Kathleen Gygi, Ph.D., Margo Hagopian, Ph.D., Richard Hannay, Ph.D., Joseph Hardy, Ph.D., Stephen Hardy, Ph.D., Gary Hartpence, Ph.D., Edward Haskell, Ph.D., S. J. Hawkins, Ph.D., Kevin Hegg, Ph.D., Lilly N. Hellman, Ph.D., Robert A. Hertz, Ph.D., Louise D. Hicks, Ph.D., Lyndon Holmes, Ph.D., Mycroft Holmes, Ph.D., O. W. Holmes, Ph.D., Tardis Hoo, Ph.D., J. E. Hoover, Ph.D., E. A. Horton, Ph.D., Lawrence Howard, Ph.D., Moe Howard, Ph.D., Ginger Hsu, Ph.D., David Hubbs, Ph.D., Loretta Huttlinger, Ph.D., Stanley Hwang, Ph.D., Harriet Kasden, Ph.D., Susan Jablonski, Ph.D., Mittie

Jackson, Ph.D., Rebecca Johnson, Ph.D., Deacon Jones, Ph.D., Edward T. T. Jones, Ph.D., Conrad Joseph, Ph.D., K. T. Kanawa, Ph.D., Liza Karpook, Ph.D., Daniel Kaye, Ph.D., William Keeler, Ph.D., Waldemar Kester, Ph.D., John M. Keynes, Ph.D., Olga Korbut, Ph.D., Susan Krock, Ph.D., Kerran Lauridson, Ph.D., Nicholas Leone, Ph.D., Meg Anne Lesser, Ph.D., Lucille S. Levesque, Ph.D., Joseph Lichtblau, Ph.D., Barbara Linden, Ph.D., Robert Lippa, Ph.D., Charles Lovejoy, Ph.D., Frances Lynch, Ph.D., Thomas Maccarone, Ph.D., Maureen Madigan, Ph.D., James Mahoney, Ph.D., Catherine Maloney, Ph.D., Jules Maigret, Ph.D., G. Maniscalco. Ph.D., Ray B. B. Mancini, Ph.D., Julius Marx, Ph.D., Cynthia Mason, Ph.D., James Matoh, Ph.D., Abigail Mays, Ph.D., Zachariah Mays, Ph.D., Charles McCarthy, Ph.D., Joseph McCarthy, Ph.D., Ann McKechnie, Ph.D., Charles Augustus Milverton, Ph.D., Robert Mishkin, Ph.D., Jack Moran, Ph.D., Charles Morgan, Ph.D., Stephen Mosher, Ph.D., Lisa Mullins, Ph.D., Sarah Natale, Ph.D., Ned Newton, Ph.D., R. M. Nixon, Ph.D., Grover Norquist, Ph.D., Ngai Ng, Ph.D., Kevin O'Malley, Ph.D., Joel Orloff, Ph.D., Frank Patterson, Ph.D., John Pesky, Ph.D., Peter Pienar, Ph.D., Margaret Pinette, Ph.D., Philip Ravino, Ph.D., Celia Reber, Ph.D., Bertrand Roger, Ph.D., Frederick Rogers, Ph.D., Dexter Rosenbloom, Ph.D., George H. Ruth, Ph.D., Kathleen Rutherford, Ph.D., Robert Ryder, Ph.D., George Scheinman, Ph.D., Aimee Semple, Ph.D., William Shoemaker, Ph.D., Joseph Slavsky, Ph.D., Olivia Smith, Ph.D., Simon Silver, Ph.D., Orenthal J. Simpson, Ph.D., Jeffrey Spaulding, Ph.D., Richard Starkey, Ph.D., David Alan Steele, Ph.D., Y. Struchkov, Ph.D., Quentin Sullivan, Ph.D., Ann Sussman, Ph.D., Ezra Tamsky, Ph.D., Kumiko Terezawa, Ph.D., Marge Thatcher, Ph. D., Mark Theissen, Ph.D., Marilyn Tucker, Ph.D., Christina Turner, Ph.D., Brenda C. W. Twersky, Ph.D., Frederick A. Von Stade, Ph.D., F. Skiddy Von Stade, Ph.D., Bertha Vanation, Ph.D., William Veeke, Ph.D., Norma Verrill, Ph.D., Y. Y. Vlahos, Ph.D., Marko Vukcic, Ph.D., Paul Waggoner, Ph.D., Teresa Wallace, Ph.D., Thomas Waller, Ph.D., J. Ward, Ph.D., John H. Watson, M.D., Michael Weddle, Ph.D., Merton Weinberg, Ph.D., Lawrence Welk, Ph.D., Kevin White, Ph.D., Andrew Williams, Ph.D., John Williams, Ph.D., Theodore Williams, Ph.D., William Williams, Ph.D., Eileen Wynn, Ph.D., Chin-chin Yeh, Ph.D., and Ethel Youngman, Ph.D.

So far as we can determine, peanut butter has no effect on the rotation of the earth.

Mondocentrism

by George Englebretsen

Philosophy Department, Bishop's University
Lennoxville, Quebec

This appeared in *AIR* 2:2 (March/April 1996).

The National Committee for the Promotion of Mondocentrism was established in 1992 to give voice to all those who believe that Mother Earth deserves a more central place in the scheme of things. In 1994 I was appointed by the Committee to lead a team of experts from a variety of scientific disciplines in a research project aimed at establishing once and for all that the Earth is indeed the center of the universe.

Infighting and Backstabbing

Our task has been hampered by much bitter infighting and backstabbing. This was unfortunate, but was to be expected from a research team consisting of astronomers, sociologists and a poet. Nevertheless, we have been able to establish conclusively that Copernicus was wrong, that the earth is indeed the center of the universe. The Copernican revolution and the subsequent history of science in the West have constituted a cruel hoax perpetuated by cynical, insensitive scientists and their flunkies.

Good Feeling is Good Policy, and That's Good

Social and political considerations (the poet was drunk and the two astronomers only work at night) have led the remnants of our team to conclude that the results of adopting mondocentrism as the official position of the United States government would have beneficial effects on both the economy and education. The first would follow from the immediate dismantling of NASA, saving literally billions of dollars in wasted resources. The second, and perhaps most important effect, would result from the fact that American students would begin to have an edge on students from other countries when it comes to performance in math and science.

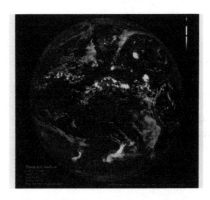

Photo: NASA Goddard Laboratory for Atmospheres.

We can prove that much of the blame for poor science performance by American students relative to foreign students must go to the fact that by placing the Earth, their home, so far from the center of everything, scientists and science educators have instilled in students a sense of low self-esteem. We can now show conclusively that when students feel good about their home, they feel good about themselves. And when they feel good about themselves, they care more about learning. If the world is at the center of the universe, then so are we. What a good feeling.

Astronomers Tend to be Misguided

One of the astronomers on our team tried to claim that we were not really scientists. (Both of the astronomers were downright snooty, with their noses up in the air all the time.) He said we should at least try to come up with some way to determine a wide variety of astronomical periodicities (or some such mumbo-jumbo) without reverting to the old cycles and epicycles of ancient, mondocentric astronomy. We eventually convinced him that once all Americans feel good about the Earth and themselves, then we won't care about the rest of the universe.

81

The Correlation Between Tornadoes and Trailer Homes

by Frank Wu

University of Wisconsin
Madison, Wisconsin

This appeared in *AIR* 1:4 (July/August 1995).

Some states are more prone to victimization by tornadoes and hurricanes than others. Expert meteorologists contend that Kansas, for example, is so often buffeted by tornadoes because of the alignment of the Rocky Mountains, the flatness of the great plains and direction of prevailing winds.[1] They are wrong. Their analysis is incomplete; it does not consider the popular notion that tornadoes are most common, simply, in states with many mobile homes[2] (see Figure 1). Below I attempt to justify and, if possible, quantify this and other common beliefs about tornadoes.

Figure 1: A mobile view of a mobile home park. Photo courtesy of the F. Wu Mobile Home Archive.

Method

A state-by-state comparison was made between the number of tornadoes and hurricanes[3] and purchases of mobile homes.[4]

Tornadoes and Trailer Homes

Data sets were standardized to each state's area[5] and results shown in Tables 1 and 2. Tornadoes and hurricanes are indeed most frequent in states with many mobile homes. For instance, eight states are in the top eleven for both prefab homes and tornadoes. Furthermore, Florida leads the nation in violent storms and is third in manufactured home purchases. Indiana ranks second in tornadoes and first in mobile homes. (It is unclear, though, what additional effects the Indianapolis 500 has on atmospheric conditions. This event, in which automobiles travel round and round at a high speed in a small oval, coincides with the beginning of the annual tornado season.)

Irrelevance of Large Cities

However good the correlation between the presence of trailer parks and tornadoes, it is not perfect. To account for the imperfections, I analyzed another popular myth,[8] the idea that states with many large cities have fewer tornadoes—it is claimed that the buildings act as windblocks. To test this idea, I divided the land area of each state[5] by the number

Table 1

1. Indiana	8.13
2. Georgia	5.43
3. Florida	4.65
4. N. Carolina	4.41
5. Alabama	4.11
6. Pennsylvania	2.91
7. Texas	2.16
8. Mississippi	1.96
9. Kansas	1.72
10. California	1.31

Table 1: Mobile home purchases per square mile. These are the American states with the highest aggregate mobile home purchases, measured in units of $1000 per square mile. Data is believed to be from 1982. Details may be contained in one or more of the works listed in the "References" section. No claims of accuracy are made or implied.

Table 2

State	Tornadoes and hurricanes (annual average)	Rural land area (in thousands of square miles)
1. Florida	7.51	27.1
2. Indiana	5.51	35.9
3. Kansas	5.10	81.8
4. Mississippi	4.82	infinite[12]
5. Texas	4.68	37.4
6. Nebraska	4.66	77.7
7. Alabama	4.07	50.8
8. Wisconsin[13]	3.44	54.4
9. Georgia	1.72	2.40
10. Ohio	1.31	3.40

Table 2: Occurrence of Violent Storms Compared with Rural Land Area. These are the American states with the highest number of tornadoes and hurricanes per square mile. Storms counts are averages for the years 1953–1990, more or less. Land area was measured in units of 1000 square miles, averaged for the years 1987–1995. Details may vary in accuracy or level of interest.

of large cities (those having a population of 250,000 or more—see reference 5); see Table 2. The occurrence of tornadoes seems to depend on both the number of mobile homes and absence of large cities. For instance, Kansas, with a whopping 81,800 rural square miles, moves up from ninth on the mobile home list to third on the tornado list. Nebraska, with 77,700 rural square miles, moves from Number 16 on the mobile home list to the sixth position on the tornado list. In contrast, California has many large cities (only 19,500 square miles between cities) and is subject to very few tornadoes (0.25 tornadoes/10,000 square mile). California, of course, makes up for the deficit in violent storms with a proud abundance of earthquakes.

Curiously, though, several large states, such as Alaska, Montana and Wyoming, have large rural areas and very few tornadoes. This could reflect either an absence of tornado-targeted trailer parks, or the absence of tornado spotters, or perhaps some other explanation.

Tornadoes and Camcorder Sales

A recent addition to tornado lore is the notion that rising sales of video camcorders somehow increased the number of tornadoes,[9] as if, perhaps, tornadoes were "posing" for pictures. To test this far-fetched hypothesis, I compared the sales of camcorders[10] to the occurrence of recent tornadoes,[11] and found that, surprisingly, there *is* a direct correlation (see Figure 2, which is omitted).

Escapable Conclusions

1. Real statistics can be used to verify virtually any hairbrained fable about tornadoes.

2. In case one of these ideas turns out to be right, I recommend that if you build a house, make it sturdy and site it near a city, far from trailer homes, behind a big rock, and don't let anyone with a video camera near the place.

References

1. *New York Times*, April 28, 1991, p. 22.

2. See: Cecil Adams, "The Straight Dope," In *Isthmus* (Madison, Wisconsin), January 14, 1994, p. 30.

3. *Storm Data,* Volume 32, Number 12 (December 1990), National Climactic Data Center, Asheville, N.C., p. I-12. Only states for which there was corresponding mobile home data are included.

4. *1987 Census of Manufactures, Industry Series, Wood Buildings and Mobile Homes*, Industries 2451 and 2452, Dept. of Commerce, Bureau of the Census, p. 24D-10. Unfortunately, some states were not included in this published data.

5. *World Almanac and Book of Facts (1991)*, Pharos Books, New York, pp. 619–43.

6. Reference omitted by common consent.

7. Reference omitted for unstated reasons.

8. Cecil Adams, "The Straight Dope," in *Isthmus* (Madison, Wisconsin), January 14, 1994, p. 30.

9. *The Capital Times* (Madison, Wisconsin), July 7, 1994, p. 3A.

10. "Annual Statistical and Marketing Reports," *Dealerscope Merchandising*, May, 1994, p. 33; March, 1992, p. 27.

11. *Storm Data*, vol. 34, no. 12, December, 1992, p. 92.

12. Infinite.

13. Wisconsin is noted for its cheese.

Low Probability of Any Further Abductions by Aliens

by Leonard X. Finegold

Physics Department
Drexel University
Philadelphia, Pennsylvania

This appeared in *AIR* 1:2 (March/April 1995).

You may well be worried about being abducted by Aliens.[1] Jacobs reports[2] on a series of detailed hypnotism-induced regression interviews (in which subjects recall their pasts) he conducted. The subjects claim to have been abducted by Aliens into UFO's (Unidentified Flying Objects).

A numerical analysis of the subjects' birthdates shows that only people born before 1970 are abducted. Hence, one may safely assume that the probability of further Alien Abductions is very low, perhaps even zero.

Analysis of the data

Jacobs gives a wonderfully simple and elegant solution to the question of why the different claimants presented such similar reports: the claimants have indeed been abducted by Aliens from UFO's.[3] A list of Abductees, with their year of birth, is given in Appendix B, p. 326 of Jacobs's book.

The following graph shows the number of Abductees in each five-year interval versus the year of birth of the Abductees (in the graph, "1965" covers the five years 1965–1969, etc.).

Some technical comments

The hypnotic regression sessions commenced in 1985 and the book was published in 1992, so the sessions were completed in about six years. (Some Abductees were not yet teenagers when abducted; one was only six years old.) This sampling time is short enough compared with the 35-year range of birthdates reported, so the zero incidence of abductions after 1969 is real. Unfortunately, Jacobs does not seem to cite the original archival journals in which the research was first reported; the index seems to be missing in my copy of Jacobs's book.

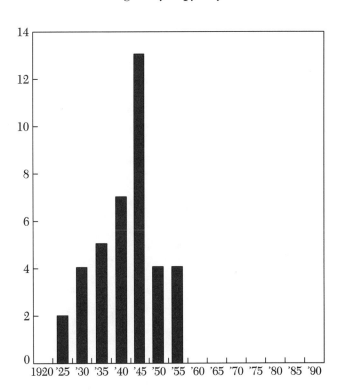

Number of abductees per half-decade (vertical scale). The horizontal scale represents the years in which abductees were born.

85

The important point to note is the decline in abductions, after the peak in abductions for birth-years 1930–1965, to zero for those born after 1969. Hence, for whatever reasons, UFO abductions have ceased in the study group for those born after 1969.[4] Since the population of the USA was expanding almost exponentially during this time, the number of abductions per 100,000 of population declined even faster than is shown in the above graph.

Conclusion

We may therefore safely extrapolate to the conclusion that the danger of further abductions by UFO's is mercifully a thing of the past.[5]

References

1. For an analysis of the legal hazards involved, see "*AIR*head Legal Review" in this issue of *AIR*. As an inducement for you to get out and journey to the library, that report is not included in this book.
2. *Secret Life: Firsthand Accounts of UFO Abductions*, D. M. Jacobs, Simon & Schuster, New York, 1992. The thesis of the book is that the Aliens temporarily abduct the Abductees into UFOs to use them in sexual reproductive processes. The book features a foreword by John Mack. (This work deservedly earned Drs. Jacobs and Mack the 1993 Ig Nobel Prize in Psychology.) Jacobs is on the faculty of Temple University. Mack is on the faculty of Harvard Medical School.
3. It should be stressed that "Aliens" means not simply "un-American," "un-French," or whatever, but "un-Earthly." Well-documented abductions by aliens from other countries are therefore excluded from this discussion.
4. The thesis of the book[3] is that the Aliens temporarily abduct the Abductees into UFO's to use them in sexual reproductive processes. I happen to have been a colleague of Edward Condon during the Condon Report work on UFO's (referred to in [3]) and was consulted during the UFO investigations (as a neighborhood friendly solid-state physicist). I can honestly aver that Jacobs' book tells me much about Aliens and UFO's of which I was not only previously completely unaware, but had no inkling whatsoever.
5. Personally I can happily report, with great relief, that my birthdate puts me in a group with zero risk of abduction.

Scientific Gossip

Contains 100% gossip from concentrate

compiled by Stephen Drew
AIR staff

These results are collected from various issues of *AIR*.

High Trash

NASA administrators, growing ever more creative as their budgets are slashed, have devised a new way to raise funds. NASA has a nearly complete catalog of the more than 10,000 pieces of man-made debris left in orbit. For a fee of $150 ("We chose the figure because it sounds reasonable," one adminis-

trator told us), NASA will allow an individual to name and own a piece of orbital debris. For an additional fee, NASA observation satellites will videotape the object's eventual fiery entry into the atmosphere. For a further charge, surviving objects can be tracked, retrieved, and delivered to the owner's home. A complete pricing schedule will be announced next month.

May We Recommend

Items that merit a trip to the library

compiled by Stephen Drew
AIR staff

These results are collected from various issues of *AIR*.

Results That Stick

Deterministic Chaos in Failure Dynamics: Dynamics of Peeling of Adhesive Tape," by Daniel C. Hong and Su Yue, *Physical Review Letters*, vol. 74, 1995, pp. 254–7. *(Thanks to investigator Claude Dion for bringing this and the next item to our attention.)*

Sheet in the Air

"Behavior of a Falling Paper," by Yoshihiro Tanabe and Kunihiko Kaneko, *Physical Review Letters*, vol. 73, no. 10, 1994, pp. 1372–5.

No Hot-Head

"The cooling power of the pigeon head," Robert St. Laurent, and Jacques Larochelle, *Journal of Experimental Biology*, vol. 194, 1994, pp. 329–39. *(Thanks to investigator Zen Faulkes for bringing this to our attention.)*

In Which the The Fix is In

"Correcting the incorrigible," G. Von Hilsheimer, W. Philpott, W. Buckley, and S. C. Klotz, *American Laboratory*, vol. 107, 1977. *(Thanks to investigator Kevin Devine for bringing this to our attention.)*

Electric Balls in a Tub

"Laboratory-Produced Ball Lightning," Robert K. Golka Jr., *Journal of Geophysical Research*, vol. 99, no. D5, May 20, 1994, pp. 679–81. *(Thanks to investigator Dahv Kliner for bringing this to our attention.)* This article desribes the author's attempts to produce ball lightning in his home by shorting the output of a 150,000 watt 10,000 amp transformer in a tub of water. The resultant fireballs:

> sizzle and hiss and skim around on the water's surface. . . . I have even seen some irregularly shaped fireballs take to the air. . . . These luminous fireballs sometimes dance right out of the tank onto the floor . . .

We welcome your suggestions for this column. Please enclose the full citation (no abbreviations!) and a photocopy of the paper.

AIR Vents

Exhalations from our readers

Note: These letters are collected from various issues of *AIR*. The opinions expressed here represent the opinions of the authors and do not necessarily represent the opinions of those who hold other opinions.

Bold Man of Late Vision

In Pelbröem Thalim's "211 Fun Facts About the Angstrom" it is claimed that "Archimedes was in effect the first person to visualize the universe from the perspective of a quark." That is not correct. I was the first person to visualize the universe from the perspective of a quark. It occurred in 1962, shortly after the invention of the quark, when my wife and I read about it in the newspaper—the Daily Telegraph—as she was preparing my coffee.

Dr. Lyle V. Todpreuss
Brighton, England

Nucular Physics

The raging arguments for the correct pronunciation of quark miss the most important principle: Murphy's law of rhyme and reason in English. Why should quark rhyme with park if work does not rhyme with pork? All arguments in the quark park are equally applicable to pork work where they do not work. Should quark rhyme with both park and pork and mean constituent quark in one case and current quark in the other? Only Murphy knows.

The obvious answer is to forget this rhyme nonsense and look at the beginnings of words where everything is clear. The quar in quark is pronounced like the quar in quart, quarrel, quarter and quarry. The wor in work is pronounced like the wor in word, world, worm, worry and even worse. All quar's are pronounced the same; all wor's are pronounced the same. Problem solved.

We leave Murphy at this point, and remember that in Brooklyn he is pronounced Moiphy. And of course Murphy's law was not really invented by Murphy, but by someone else with the same name.

Harry Lipkin
Weizmann Institute
Rehovot, Israel

CHAPTER 5

The New Chemistry

A few years ago, an American television advertisement explained that:

WITHOUT CHEMISTRY, LIFE ITSELF WOULD BE IMPOSSIBLE

We at *AIR* have always been jealous of whoever wrote that ad. It says everything, and it says nothing. It is fatuous. It is perfect.

Anyway, without chemistry, this chapter would be impossible, or at least pointless. Chemistry seems to intimidate people. Bubbling test tubes, obscure Germanicallystrungtogetherlonglonglongchemicalnameswithnumbersinthem4godknowswhatreason, rumors that those who go into the profession have higher mortality rates than anyone else—that is chemistry as viewed by much of the general public. That and the phrase "organic chemistry," which for some (fairly good) reason strikes terror into the heart of the tender pre-med student.

But chemistry really isn't scary, and chemists do know how to have a good time, and occasionally they even go out and have that good time. Moreover, chemistry is exceedingly useful to the common person. Look at Scott Sandford's "Apples and Oranges—A Spectrographic Comparison." Sandford provides you with a simple, devastating weapon to use the next time someone accuses you of "comparing apples and oranges."

Scientist/supermodel Symmetra uses her "Ask Symmetra" column to solve people's personal problems. Symmetra has been described as "Ann Landers with modeling experience and a knowledge of advanced chemistry." The description is not without merit.

The long-running controversy about how to measure human intelligence has finally been solved. Chemist Dudley Herschbach does what Freud and the *Bell Curve* boys, Richard Herrnstein and Charles Murray, authors of the strange book of that name, could not. He brings quantum physico-chemical principles to bear on the problem. You can read about it in "Quantum Interpretation of the Intelligence Quotient (QI of IQ)."

Alice Shirrell Kaswell keeps a close watch on obscure research journals. In "Cindy Crawford Discovers," Kaswell presents several key discoveries that were almost lost amidst today's overwhelming profusion of scientific journalism.

Our high school readers will find two handy study aids here, Robert Rose's "Politically Correct Periodic Table of the Elements" and a scratch n' sniff poster that will horrify teachers.

Anyone who is concerned at the rising cost of lab equipment should be delighted with David Cann and Phillip Pruna's report on how to do "Xerox Enlargement Micrography." Cann and Pruna have made obsolete the quaint electron microscopes of old.

Apples and Oranges: A Comparison

by Scott A. Sandford

NASA/Ames Research Center
Moffett Field, California

This appeared in *AIR* 1:3 (May/June 1995).

We have all been present at discussions (or arguments) in which one of the combatants attempts to clarify or strengthen a point by comparing the subject at hand with another item or situation more familiar to the audience or opponent. More often than not, this stratagem instantly results in the protest that "you're comparing apples and oranges!" This is generally perceived as a telling blow to the analogy, since it is generally understood that apples and oranges cannot be compared.

However, after being the recipient of just such an accusation, it occurred to me that there are several problems with dismissing analogies with the comparing apples and oranges defense.

First, the statement that something is like comparing apples and oranges is a kind of analogy itself. That is, denigrating an analogy by accusing it of comparing apples and oranges is, in and of itself, comparing apples and oranges. More important, it is not difficult to demonstrate that apples and oranges can, in fact, be compared (see Figure 1).

Materials and Methods

Figure 2 shows a comparison of the 4000–400 cm^{-1} (2.5–25 μm) infrared transmission spectra of a Granny Smith Apple and a Sunkist Navel Orange.

Both samples were prepared by gently desiccating them in a convection oven at low temperature over the course of several days. The dried samples were then mixed with potassium bromide and ground

Figure 1: A Granny Smith Apple and a Sunkist Navel Orange.

in a small ball-bearing mill for two minutes. One hundred milligrams of each of the resulting powders were then pressed into a circular pellet having a diameter of 1 cm and a thickness of approximately 1 mm. Spectra were taken at a resolution of 1 cm^{-1} using a Nicolet 740 FTIR spectrometer.

Conclusions

Not only was this comparison easy to make, but it is apparent from the figure that apples and oranges are very similar.

Thus, it would appear that the comparing apples and oranges defense should no longer be considered valid. This is a somewhat startling revelation. It can be anticipated to have a dramatic effect on the strategies used in arguments and discussions in the future.

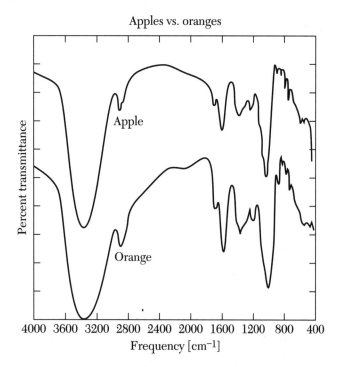

Apples vs. oranges

Figure 2: A comparison of the 4000–400 cm⁻¹ (2.5–25 5 μm) infrared transmission spectra of a Granny Smith Apple and a Sunkist Navel Orange.

A Personal Note

I, for one, intend to carry a copy of Figure 2 in my pocket so that the next time someone accuses me of comparing apples and oranges, I can whip it out, shove it in front of their eyes, and say, "No—*this* is comparing apples and oranges!" That should fix them.

Ig Nobelliana
Words for the ages

"I'd be pleased to have [Edward] Teller get a second Ig Nobel Prize so he could become listed in the Guinness Book of Records as the person who's achieved the most Ig Nobel Prizes."

—*Linus Pauling, winner of the Nobel Prize in Chemistry (1954) and the Nobel Peace Prize (1962), and founding editorial board member of The Annals of Improbable Research.*

Xerox Enlargement Microscopy (XEM)

by David P. Cann and Phillip Pruna

Materials Research Laboratory
Pennsylvania State University
University Park, Pennsylvania

This appeared in *AIR* 1:2 (March/April 1995).

A revolutionary new microscopy technique makes it possible to achieve subatomic resolution levels with standard copying machines. In the past, high resolution had been achieved through existing techniques such as transmission electron microscopy (TEM), atomic force microscopy (AFM), etc. A full-scale revolution in thinking was required to break the existing limits set by these archaic methods. The authors present Xerox Enlargement Microscopy (XEM), a technique that will take high resolution microscopy to another new and exciting realm. See Figure 1.

Description of the Technique

There are a number of important advantages to this new technique. First and foremost, it is an extremely simple technique. Figure 2 describes the procedure in flow chart form. Since copy machines are already in existence in most laboratories, no new expenses are required. Most machines can be operated at a cost of approximately five cents per page, significantly less than the current rates for operating a TEM or STM.

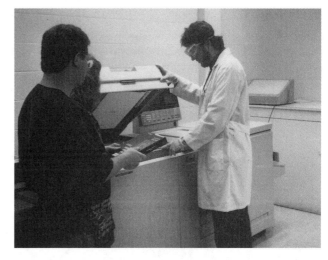

Figure 1: Technicians perform Xerox Enlargement Microscopy, using standard copying machines.

No sample preparation is necessary at all. Figure 3 is an XEM micrograph of ferroelectric barium titanate ($BaTiO_3$) magnified 15,392 X. This micrograph was taken using a Xerox 1090 copy machine in collation/stapling mode on BaTiO3 in powder

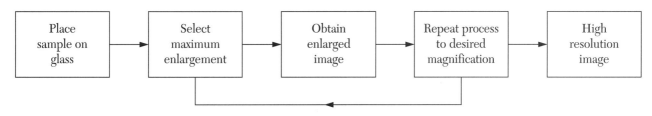

Figure 2: Flow chart depicting the XEM experimental procedure.

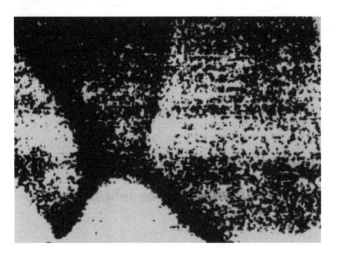

Figure 3: XEM micrograph of BaTiO₃ magnified 15,392 X.

form. For the Xerox 1090, the maximum enlargement is 155%, so that it required 22 enlargement steps to achieve 15,392 (1.55^{22} = 15,392).

Collated and Stapled XEM

On the more sophisticated XEM instruments, a collation and/or staple option may be available. This is a powerful tool which, to the knowledge of the authors, has no parallel in the other high resolution imaging techniques.

Ultra High Resolution XEM: Images of Atomic Hydrogen

By enlarging 48 times, an incredible magnification of 1,367,481 X was achieved on samples of deuterated ammonium hydrophosphate ($NH_4H_2PO_4$). For the first time, a single deuterium ion could be imaged, as seen in Figure 4. One can see remarkable evidence

for the Heisenberg uncertainty principle as we see the "quantum fuzziness" of the proton-neutron nucleus and orbital electron.

Conclusions/Future Work

A simple, cost effective, high resolution technique has been presented. Further work currently under investigation is divided into two main areas. First, the possibility of obtaining diffraction data from the XEM images is being investigated by theoreticians. Secondly, our experimentalists are attempting to probe the nucleus via XEM and verify the existence of quarks.

References

1. *Opticks*, Isaac Newton, 1704.
2. *Xerox 1090 Operation Manual*.
3. Mongolian Patent number 4, 1993.
4. Private communication with Dr. Clive A. Randall.

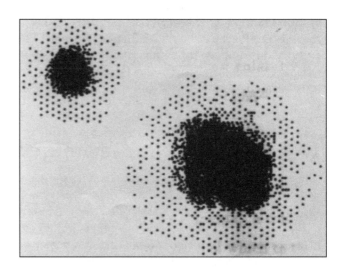

Figure 4: XEM micrograph of a deuterium atom.

Ask Symmetra

Elegant solutions to complex problems

by scientist/supermodel Symmetra
AIR staff

The items here are collected from several different issues of *AIR*.

My fiancée is desperate to find a color that will go well with her eyes, which are a very deep brown. Our wedding is next month, and we are expecting unusually warm weather. Can you suggest something that's not too stodgy?

— *A.O*

Why not try something like Prussian blue? It has a tinkertoy molecular structure with $M \leftarrow C \equiv N \rightarrow M'$ linkages in 3-D. Of course the choice of metals and their oxidation states affect the magnetic behavior. If you use Ni^{II} and Cr^{III}, which is to say that you have $CsNi[Cr(CN)_6].2H_2O$, the material is a ferromagnet with a critical temperature of 90K. Go for it. Her relatives will approve, your friends will be pleased, and the marriage will get off to a good start.

Dear Symmetra:

My spouse's kitchen habits are threatening my marriage. I love to cook special meals for her, but then she keeps the leftovers in the refrigerator so long that I fear they will either disintegrate or explode. Can this marriage be saved?

— *P.F.*

Food is the subject of as many marital arguments as money, sex and in-laws. I wouldn't worry about the leftovers disintegrating—it happens to me all the time. As for explosions, the most relevant performance parameters are detonation velocity and pressure. The equations are:

$$\text{detonation velocity} = 1.01\sqrt{\{NM^{1/2}Q^{1/2}(1+1.3\rho_0)\}}$$
$$\text{detonation pressure} = 15.58(\rho_0)^2 NM^{1/2}Q^{1/2}$$

where N = the number of moles of gas per gram of explosive, M = the average molecular weight of the gases, $Q = \Delta H_0/gram$, ρ_0 = initial density, and ΔH_0 is the free energy of decomposition to give CO, N_2 and H_2O (g) at standard temperature and pressure. The bigger question is why neither of you wants to eat those leftovers. I suggest a marriage counselor—either that or eat out more often.

Dear Symmetra:

My brand new significant other just moved in with me. This incredible soulmate loves the tender, silky feel of my hair but not the color, and wants me to dye it. I've never done anything like this. Can you tell me what's involved?

—*N.P.E.*

There are two aspects to this. First, living with someone is a whole different experience than simply dating. Second, the basic reaction involving hydrogen peroxide is rather unstable:

$$2H_2O_2 \rightarrow 2H_2O + O_2(g)$$

The decomposition is slow, but it's catalyzed by dust, dissolved compounds and other impurities. The reaction also is accelerated in the presence of light, which is why you generally see it stored in dark bottles. You might want to keep that in mind.

I think that if you can get past the logistics of who uses what in the bathroom, you've got a shot at making it last. I hope it works out!

If you have a question for scientist/supermodel Symmetra, send it to: "Ask Symmetra," The Annals of Improbable Research, PO Box 380853, Cambridge, MA 02238 USA. Because of volume of mail, replies to personal invitations are not possible.

Science Demonstration: Scratch 'n' Smell

For beginning and intermediate chemistry students

by LaDuc Foment
AIR staff

This appeared in 1993.

This special handout was prepared using microencapsulation techniques developed at 3M Corporation.

Scratch here to smell the characteristic odor of the chemical substance

H_2O

Scratch here to smell the characteristic odor of the chemical substance

O_2

Quantum Interpretation of the Intelligence Quotient (QI of IQ)

by Dudley Herschbach
Nobel Laureate (Chemistry, 1986)
Chemistry Department, Harvard University

This appeared in *AIR* 1:1 (January/February 1995).

The interpretation of IQ scores has been notoriously contentious for 80 years.[1] Moreover, explanations are entirely lacking for some striking observations, such as the recent discovery[2] that listening to music by Mozart temporarily raises IQ significantly, by nearly 10 points. Here I outline a new interpretation derived from quantum physics, the "QI of IQ."

My basic hypothesis is that intelligence arises from the vibrating molecules within our brains. There are a great variety of molecules, so our brains certainly oscillate with a wide range of frequencies. However, for simplicity, I adopt the approximation employed by Einstein in his famous paper of 1907 treating the heat capacity of solids.[3] This represents the net effect by a single harmonic oscillator. The vibrational frequency, F, is proportional to the square root of the ratio K/M, where the constant K denotes the stiffness of the vibrating tissue (varying from airy to rocky) and M the effective mass (ranging from light- to heavy-headed). For the present qualitative discussion, IQ is considered to be directly proportional to the vibrational amplitude, although the precise relationship will need to be determined experimentally.

Figure 1 shows the probability distributions of the amplitude for the lowest three allowed quantum states of an oscillator, labeled by quantum numbers n = 0,1,2. Both the positions and number of undulatory maxima differ markedly for these distributions. The relative populations of the various states depend on the ratio F/T, where T is an effective temperature governed by interactions with the environment.

If T is sufficiently low compared to F, most of the oscillating molecules reside in the lowest energy state, n = 0, termed the ground state. The probability distribution of its vibrational amplitude indeed has exactly the same bell-shaped form as the IQ-distribution curve. According to the conventional IQ scale, the peak corresponds to IQ = 100. Scores one standard deviation higher or lower, IQ = 115 or 85, respectively, correspond to the maximum and minimum vibrational amplitudes within the region allowed by classical mechanics (within the dashed parabolic curve). Amplitudes beyond this region, although classically forbidden, are allowed by quantum mechanics, but with rapidly decreasing probability. This is a consequence of the celebrated "tunnel effect," which permits a quantum particle to sneak into places that it does not have enough energy to reach. In the ground state, the total probability of achieving an IQ above 115 is about 16%. For IQs above 150, usually considered the "genius" range, the total probability is only 0.04%; that corresponds to only 400 people per million.

Fortunately, such "tunneling enhancement" is not the only way to achieve high IQs. If the temperature is high compared to the characteristic frequency, the oscillating brain may visit excited states more often than it resides in the ground state. As seen in Figure 1, in an excited state, the maxima of the probability distributions shift to wider amplitudes. The classically allowed region also becomes much larger than in the ground state, by a factor of the square root of 2n + 1. By virtue of this factor alone, a ground state IQ of 115 would be amplified to 150 in an excited state with n = 5. This indicates that "temperature enhancement" as well as tunneling must have a major role in human intelligence.

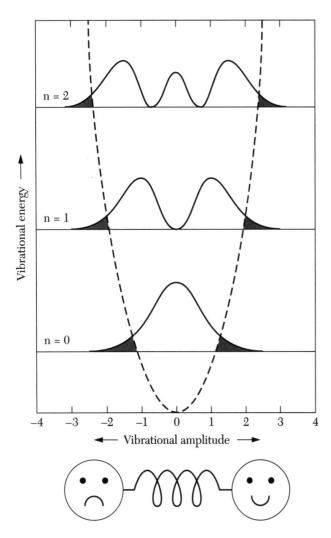

Figure 1: Probability distributions for the lowest three quantum states of a harmonic oscillator (n = 0, 1, 2). Dashed parabola indicates region of motion allowed by classical mechanics; shaded regions show domains made accessible by quantum tunnel effect. Vertical scale indicates energy in multiples of hF, where h is Planck's constant and F the oscillator frequency. Horizontal scale indicates vibrational amplitude. According to the QI of IQ, the midpoint of the horizontal scale (zero amplitude) corresponds to IQ = 100; positive and negative excursions (expansion or contraction of the oscillator) are indicated in multiples of the standard deviation, which is 15 IQ units.

Despite the undulations in the excited states, the sum of the probability distributions, with each state weighted by its population, in fact again has the same bell-shaped form as for the ground state. The width of the net distribution, however, is wider, reflecting the temperature-dependent population of the excited states.

Implications and ramifications of this QI of IQ are readily apparent, even to anyone well below the mean. Here I note only a few consequences:

1. The importance of the ratio F/T resolves the old controversy about heredity vs. environment. For somebody born with a rock-hard head (large K) or an intellectual lightweight (small M), the frequency F will be large, hence the vibrational amplitude small. Yet a sufficiently large T, supplied by a "hot" intellectual environment, can still provide a favorable F/T ratio.

2. While people subject to intellectual excitement clearly can experience a large enhancement of IQ, that can occur also in a relaxed mode. This is analogous to pushing a child on a swing; a large amplitude can be obtained either by a mighty shove or gentle nudges, properly timed. The hitherto mysterious effect of Mozart's music surely results from such gently resonant nudges.

3. Since any oscillator contracts as well as expands, both the tunneling and temperature effects operate just as strongly to depress IQ. This immediately explains a commonly observed phenomenon (shamefully ignored by psychologists ignorant of quantum dynamics), the fact that often smart people do dumb things.

References

1. *The Bell Curve*, R.J. Herrnstein and C. Murray, New York: The Free Press, 1994.
2. "Music and Spatial Task Performance," F. H. Rauscher, G. L. Shaw, and K. N. Ky, *Nature*, vol. 365, 1993, p. 611.
3. *Annalen der Physik*, A. Einstein, vol. 22, no. 180, 1907.

The Politically Correct Periodic Table

by Robert Rose

Department of Materials Science
MIT, Cambridge, Massachusetts

This appeared in 1993.

To protect the health and welfare of the general public, we must eliminate all sources of toxicity, pollution and radioactivity. Isotopes and artificial elements should not be tolerated, nor should sources of greenhouse gases or hypertension. Sexist nomenclature has no place in modern society.

These requirements can be satisfied by making the indicated revisions to the periodic table of the elements.

1 H																	2 He
3 Li	*T*											*T*	*G*	*P*	8 O	*H*	10 Ne
B	12 Mg											13 Al	14 Si	*P*	*P*	*H*	18 Ar
19 K	20 Ca	21 Sc	22 Ti	23 V	*P*	25 Mn	26 Fe	27 Co	28 Ni	29 Cu	30 Zn	31 Ga	32 Ge	*T*	*S*	*H*	36 Kr
37 Rb	38 Sr	39 Y	40 Zr	41 Nb	*S*	*R*	*S*	45 Rh*	46 Pd*	47 Ag*	*T*	49 In	50 Sn	*T*	*S*	*H*	54 Xe
55 Cs	56 Ba	57† La	72 Hf	73 Ta	74 W	75 Re	76 Os	77 Ir*	78 Pt*	79 Au*	*T*	*T*	*T*	83 Bi	*R*	*R*	*R*
R	*R*	††	*A*	*A*	*A*												

	58 Ce	59 Pr	60 Nd	61 Pm	62 Sm	63 Eu	64 Gd	65 Tb	66 Dy	67 Ho	68 Er	69 Tm	70 Yb	71 Lu
†Lanthanides														
††Actinides	*R*	*R*	*R*	*R*	*R*	*R*	*R*	*A*	*A*	*A*	*A*	*A*	*A*	*A*

N.B. No isotopes of any kind are permitted.

T – toxic element

P – pollutant

S – sexist nomenclature

B – hazardous element – raises blood pressure

G – source of greenhouse gas

H – halogens not permitted — see all of above

R – radioactive elements not permitted

A – artificial elements not permitted without prior approval

* – surtaxes on these elements

Cindy Crawford
DISCOVERS

The face value of science

by Alice Shirrell Kaswell
AIR staff

This appeared in *AIR* 1:6 (November/December 1995), and focuses on findings in the September 1995 issue of *Redbook*.

In this regular column, I summarize important scientific discoveries that were reported in the pages of obscure research journals such as *Cosmopolitan, Vogue, GQ,* and the *New York Times*. Often, the general public reads these reports but does not realize their significance. I believe that a cosmetics ad or a stock market summary is just as important and correct as some of the scientific reports that are published in *Nature* or *Science*. And they are generally much more attractive.

Crawford Combinatorics

Supermodel/actress **Cindy Crawford** has been conducting experiments with men. A probability analysis performed by investigator **Jane Heller** suggests that Crawford is not finished with men forever. A five page report about Crawford and her recent work appears in the September, 1995 issue of the research journal *Redbook*.

Calvinist Longing

Investigator **Calvin Klein** is continuing his possibly endless research on the topic of "eternity." Details—skimpy details—appear on pages 33–4 of *Redbook*. An accompanying photograph of two wet people is not properly labeled. My guess is that one of the people is Klein and the other is not.

Skin Hard as Nails

Investigator **Sally Hansen** is conducting research that involves "Skin Recovery with Alpha Hydroxy and Botanical Complex." Details appear on pages 36–7. The report concludes that "Sally Hansen [is]

the most trusted name in nails." I don't know what any of this means, but I intend to find out.

The Riches of Nutrium

I have long been an admirer, from afar, of the **Pond's Institute**. One of the few institutes with an 800 number (800-34-PONDS), it performs research with a brand of panache that has been lacking in this century. A report on page 45 tells about "Skin Smoothing Capsules" that have "instant indulgence" which can be released "with a twist and a squeeze." According to the report, the significant thing about this is that "softness is back." The Institute has also been investigating "The Riches of Nutrium," a substance of some sort which includes "vitamins, essential lipids, and an alpha hydroxy compound." An accompanying photograph shows a woman with cleavage.

Hurley's Construction

On page 47, supermodel **Elizabeth Hurley's** engineering work is described. Hurley has apparently pioneered a process (or a substance?) called "Versace." An accompanying photo shows one of Hurley's new projects, a "daring, held-together-by-safety-pins Versace dress."

Emolients, Toners and Volumizers

An unnamed *Redbook* investigator presents a summary overview of recent research efforts. The findings include the following.

1. "Women over 30 should avoid dark mascara."
2. Contouring color, whatever that is, can make cheekbones look prominent.
3. "Night creams *are* more emolient than day lotions."

4. "Most women don't need to use a toner." I have also found that most men do not know what a toner is. Nor do I, exactly.

The report also discusses volumizers. As with Sally Hansen's nails (see above), I don't know what volumizers are, but I will attempt to find out.

Cox's Chromium Picolinate

Actress **Courteney Cox** has bottles of vitamin C and chromium picolinate in a ceramic bowl. Lucky girl. Details are on page 116. I wish I had chromium picolinate. I already have a ceramic bowl.

AIR Vents

Exhalations from our readers

Note: These letters are collected from various issues of *AIR*. The opinions expressed here represent the opinions of the authors and do not necessarily represent the opinions of those who hold other opinions.

An Opinion of Difference

Scott A. Sandford ("Apples and Oranges: a Comparison") raises a very important question concerning the idiom "comparing apples and oranges." However, his conclusion has a linguistic bias. In Swedish the idiom reads, after translation, "comparing apples and pears." Indeed, apples and oranges are quite similar, as Dr. Sandford concludes, but a Granny Smith Apple and an Anjou Pear are not that similar at a visual inspection. I urge Dr. Sandford to continue with his important research.

Anders Larsson
Husbyborg, Uppsala, Sweden

Protect the Children

I do not subscribe to your journal for the type of cover on your May/June 1996 issue. One of the things I stress when I teach high school students is the importance of safety in the chemistry lab. Clearly, the two gentlemen (behind the scientist in the foreground) are not wearing proper safety goggles. How can I expect my students to follow proper OSHA standards when their role models in university chemistry labs clearly do not?

Jamie Larsen, Science Teacher
Sedona, Arizona

CHAPTER 6

Biology and Medicine

Biology is the science of mysteries. It can be really, truly, exasperatingly hard to nail down exactly how and why any particular kind of animal, plant, bacterium, organ, cell, or bodily system works. Within any individual creature, so much is going on, at so many levels, at every moment, that living things are not at all simple to figure out. They are all extremely complicated, and very varied, from individual to individual and from one stage of development to another within any one of those individuals. That's why there are so many apparently crazy study results. You can do one study on almost anything and apparently "find" all kinds of outlandish things. If a whole raft of studies find the same outlandish things, though, then maybe those findings are not so outlandish. Keep all that in mind when you read the apparently outlandish individual studies in this and the next chapter.

Biology is the study of living things. Medicine is the attempt to keep living things living. In this chapter, you will see much of the former, with some evidence of the latter.

One day in the spring of 1994, Earle Spamer of the Academy of Natural Sciences, Len Finegold of the Drexel University Physics Department, and that fellow Marc Abrahams were walking out of the studio of station WHYY in Philadelphia, where we three had spent an hour successfully distracting radio host Marty Moss-Coan while she tried, tried, tried to conduct her call-in radio show. Len kept taking flash photos of Ms. Moss-Coan and showing her his kit bag full of teaspoons, WHYY umbrellas, and knickknacks from his childhood in England. As we hastily left the station, I mentioned that I had come up with a good title for an article, but hoped that either Earle or Len would like to do the research and write it up. Thus was born "The Taxonomy of Barney," which Earle wrote over the next few weeks together with two other colleagues. Their article answers the question, "is Barney the TV dinosaur in fact a dinosaur?" The article was published in the premiere issue of *AIR*. It has brought its authors attention from scientists, parents, and

105

television stations around the world, and from adoring schoolchildren who visit the Academy of Natural Sciences to hear a talk on the taxonomy of Barney and see a glass specimen jar that contains a specimen of Barney preserved in formaldehyde.

In addition to photos of Barney, this chapter contains several of the kinds of genuine scientific photographs for which *AIR* is renowned. There are remarkable sights that people discovered in their microscopes ("Happy Yeast" and "The Surfer Girl Fungus"), and one from a South African beach ("The Sad Crab of South Africa"). Mark Benecke's article, "Nematodes and Hieroglyphs," presents a plausible explanation for how the microscopic worms called nematodes may have influenced the development of human writing. "The mickeymouse Gene," shows a photograph of a DNA gel, the sort that became known to the public during the O. J. Simpson trial.

P. A. Paskevich and T. B. Shea's "The Ability of Woodchucks to Chuck Cellulose Fiber" is an attempt to settle an age-old conundrum.

In "A Natural History of the Articulated Lorry," Angela Close applies her analytical skills to a species that other scientists had neglected.

V. D. Irby and M. S. Irby spent an entire summer watching grass grow. They want to share their excitement with you. Read, if you will, their fascinating treatise, "Cyclic Variation in Grass Growth."

As you have seen in the chapter on the Ig Nobel Prizes, Jon Marks made his mark on the world with the landmark study, "Arivederci, Aroma: An Analysis of DNA Cologne." You can read the original report right here in this chapter.

Yeast is a fascinating subject, especially when it affects human beings. You will not soon forget the lesson taught by the medical report, "A Man, a Woman, a Yeast."

This chapter also contains the very first of *AIR*'s long series of reviews of cafeterias at the world's great research institutions. Karen Hopkin's report on Blackford Hall at Cold Spring Harbor Laboratory will stimulate your scientific appetite.

The Taxonomy of Barney

Evidence of Convergence in Hominid Evolution

by Edward C. Theriot,[1,4] Arthur E. Bogan,[2,5] Earle E. Spamer[3,4]

This appeared in *AIR* 1:1 (January/February 1995).

Introduction

The evolution of hominids is a controversial subject. The fraudulent case of "Piltdown Man" has charged this area of research with wariness, and the tragic loss of the specimens of Peking Man has introduced political intrigue. Of course, one should not overlook the impassioned conflicts surrounding Creationist viewpoints of the rise of Man,[6] but we are unable to compare these data with ours, so we only present our data.

The Problem

According to *National Geographic*,[7] hominids evolved first on the African continent, radiating to occupy the other continents during the past tens to hundreds of thousands of years. Current opinions put forth by anthropologists indicate that several genera and species evolved, of which only *Homo* exists today. The only evidence on which these suppositions are based are skeletal remains, preserved mostly as fragments. Cladistic studies of the characteristics of the bone fragments have led scientists to derive the evolutionary relationships between these different hominid animals.

However, from field evidence and empirical observations, we have discovered a previously unrecognized form of hominid, alive today, which is presumably globally distributed. It is certainly found in North America, where we first observed it. Its external morphology is completely unlike hominid morphology, for which reason it has been until now overlooked. Its discovery has immediate and far-reaching implications on understanding hominid evolution.

Materials and Methods

In February 1994, we observed on television an animal that was there identified as a dinosaur, Barney.[8] Its behavioral characteristics suggested that it was dissimilar to the diverse dinosaurian faunas that are so well documented.[9] Even accounting for the probability that some dinosaurs were socially closely organized, and that some even may have been warmblooded, Barney's animated attitude, communication skills, and worshipful relationship with juvenile specimens of *Homo*, all pointed to an unrecognized aspect of reptile form and function.[10]

To test the hypothesis that Barney is a reptile descended from the true dinosaurs, we went into the field to capture and study a living specimen. This we accomplished with remarkable ease, as Barney was advertised to be appearing at a local shopping mall. In a secure area, we established an observation post, which met the immediate need for controlled documentation of Barney's external physical characteristics.

Additional instrumentation was required to determine the internal structure of Barney. We elected not to sacrifice the specimen, as we believed that this would have had a negative impact on the associated fauna (the juvenile specimens of *Homo*). Mostly noninvasive procedures were designed to obtain our data. A wide-field X-ray emitter was built to obtain images of the skeletal structure of Barney. Unexposed X-ray film plates were hung decoratively on the wall near where Barney was expected to show itself; they were not noticed by any of the human subjects, nor by mall security. The X-ray emitter had only short exposure times, thus we believe that the human subjects in proximity to Barney were in no danger greater than were the residents of Chernoble.

Our cladistic analyses were made using both PAUP and Maclade character analysis programs, but we resorted solely to the Maclade results because they print out much more nicely.

Observations

The specimen we observed is 183 cm tall, has a dinosauroid shape, is bipedal, and has a head about one-third the size of the body. Two eyes are positioned on the front of the head, suggesting binocular vision. There are no visible auditory openings. Two depressions occur on the outside extremity of either side of the snout, in position of nasal openings. The mouth contains two smooth, white structures, one affixed to the dorsal part of the oral opening, the other to the ventral part of the oral opening; they have the form and position of dental batteries. The mid-section of the body is greatly distended. The epidermis is entirely of fuzz, colored purple except on the belly where it is green; there are two spots on the posterior. The limbs are each in two major segments. The forelimbs are short and bifurcated at their extremities; there are no nails or claws. The legs are squat, with wide feet; each foot has three nails or claws but otherwise has no separate digits. A broad, cylindrical, tapering tail extends from the posterior about 100 cm; it appears to be fused, without musculature, and was not observed to move voluntarily. Barney is otherwise externally bland.

X-ray photographs of Barney have provided our most astounding observations (Figure 1). The skeleton is not that of a reptile, but it is clearly hominid both in morphometry and distribution of osteological elements. In fact, it is indistinguishable from the skeleton of *Homo*. The pelvic structure is mammalian; there are heterodont teeth with a dental formula precisely that of *Homo*; there are five digits on each of the extremities; and there are no vertebrae beyond the coccyx of the vertebral column, leaving the tail without skeletal support. However, the presence of a coelom, or body cavity, separating the skeleton from the dermal structure, makes Barney very unlike mammals and reptiles.

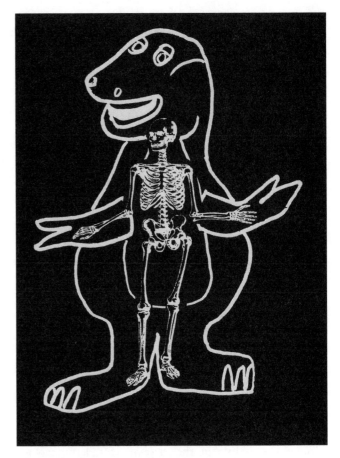

Figure 1: Composite image of Barney showing external morphology and skeletal structure.

Analysis

The external morphology of Barney belies its mammalian affinity. Evolutionarily this suggests some selective advantage, to have the external form of a dinosaurian reptile and the internal structure and abilities of a hominid mammal. This view is supported by Barney's observed ecological niche and behavioral characteristics, where it is always in association with juvenile hominids. The association seems to be one of co-dependence, and we present conjecture that Barney has evolved into the niche occupied by juvenile hominids, who by their own nature occupy a very protected part of hominid social structure, thus Barney would effectively ensure its survival by integrating itself into this environment.

This still does not explain the taxonomic relationship of Barney to other vertebrates. To examine this, we compared various physical characters of Barney with the characters of other mammals, reptiles, birds, and fish (Table 1). We selected characters based on

Table 1. Characters used to compare Barney to other vertebrates

	Barney	Mammal		Dinosaur		Bird	Salmon	
		Human	Whale	Ornithischian	Saurischian		(Live)	(Dead)
Dermal structure	fuzz	hair	hairless	scales	scales	feathers	scales	fuzz
Tooth structure	heterodont	heterodont	homodont	homodont	homodont	homodont	heterodont	heterodont
Pelvic structure	mammalian	mammalian	mammalian	ornithischian	saurischian	ornithischian	ichthyschian	ichthyschian
Claws/nails	yes	yes	no	yes	?	yes	no	no
Fusion of extremities	yes	no	yes	no	no	no	yes	yes
No. of leg segments	2	2	0	3	3	3	0	0
Coelom	yes	no	no	no	no	no	no	yes
Oral display	yes	no	no	no	no	no	no	yes
Tail	yes	no	yes	yes	yes	no	yes	yes
Mammary glands	no	yes	yes	?	?	no	no	no
Lungs	yes	yes	yes	yes	yes	yes	no	no
Live birth	no	yes	yes	no	no	no	no	non-productive
Binocular vision	yes	yes	no	no	no	no	no	no vision
Blood	warm	warm	warm	?	?	warm	cold	cold or gel

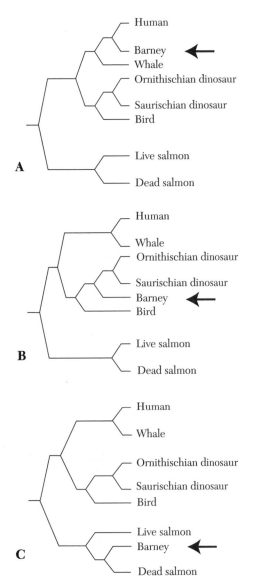

Figure 2: Derived trees comparing Barney to A) humans (the most parsimonious tree, 29 steps), B) dinosaurs (32 steps), and C) salmon (31 steps).

were most interested in the results of the tree in which Barney was grouped with them. This less parsimonious tree contains 32 steps (Figure 2b). We compared Barney to the outgroups of live and dead salmon. We correctly predicted that Barney was very unlike a live salmon, but we were very surprised to find that the tree comparing Barney to a dead salmon (Figure 3) was more parsimonious (31 steps, Figure 2c) even than the tree which grouped Barney with the dinosaurs.

The remarkable similarity of Barney to dead salmon emphasizes the distinctly non-reptilian characters. In each, the dermal covering is fuzz, a coelom is present, and an oral display character is present and independent from the dentition. This last character is of particular note. In Barney, the oral display (see Figures 1 and 6) appears to serve no active function. There is no similar feature among the reptiles. This non-functional display is similar to the terminal sexual display character of the salmon. However, since Barney appears not to be in a reproductive mode, we have compared the oral display to one of territorial

their affinities across the spectrum of vertebrates. We added or discarded characters until we achieved the results we believed, then stopped. Barney was compared to humans, whales, ornithischian and saurischian dinosaurs, and birds. In the cladistic diagrams our outgroups are live and dead salmon.

Selected characters were scored and run through the Maclade program. We first derived the most parsimonious tree, in which Barney was shown to be most similar to humans; the tree has 29 steps (Figure 2a). Then Barney was grouped with the other vertebrates to determine how many steps the trees produced. Since Barney has external morphometric affinities to the bipedal ornithischian dinosaurs, we

Figure 3: A dead salmon with hominids. Note that Barney more closely resembles the hominids and the dead salmon than it does the dinosaurs (not shown).

Figure 4: Example of territorial display in hominids.

demarkation. We have observed similar means of territorial display in hominids (Figure 4), which again reinforces Barney's affinity to the Hominidae rather than the Reptilia.

The very animated, social behavior we observed in Barney also indicates an affinity to hominids. The behavior suggests warm-bloodedness, which we sought to document in the specimen we studied. This required a temporarily invasive procedure, which we performed when the subject was found alone in a hallway. The subject was uncooperative and escaped, thus our measurements of Barney's body temperature are inconclusive. We suspect that our failure to com-

Figure 5: Instrumentation used to establish blood temperature of Barney.

plete the procedure was due to inadequate instrumentation (Figure 5).

Implications for Evolution

We have demonstrated that Barney is most similar to humans. Yet it is more like a dead salmon than even the dinosaurs to which group it purports to belong! We interpret this to be a case of convergence in evolution, where the ancestral Barney has evolved to occupy the same ecological niche as that now containing juvenile hominids.

Figure 6: A specimen of Pretendosaurus barneyi.

This poses significant questions to the interpretation of the fossil record. Non-skeletal materials are rarely preserved as fossils. It is therefore likely that the only part of the Barney animal to be found as a fossil is its skeleton, and we raise the question of misidentification of fossil remains. The criteria hitherto used to identify the skeletons of early humans and their precursors are non-indicative. If a skeleton of a proto-human cannot be distinguished from that of Barney, there is a likelihood that some of the skeletal specimens of early hominids—"Lucy" for example—may in fact be the skeleton of a Barney ancestor.

Conclusion

Barney is not a dinosaur. It is a hitherto unknown member of the Family Hominidae, which we name *Pretendosaurus barneyi* (from the Latin, *pretendo*, meaning "allege, simulate, pretend, or pretender," and *saurus*, "lizard"). Its fossil record is presently unknown, but we infer from our data that it may extend to the Early Paleolithic Era. A complete reexamination of fossils said to be ancestors of humans is called for. The cultural cliché of coexistence of dinosaurs and humans, so richly represented in film (e.g. *King Kong* and *The Flintstones*), similarly may benefit from reexamination in light of the evidence seen in Barney, from which some significant sociological and anthropological conclusions may be derived.

That Barney can be sighted today in numerous places is a sure indication of a widespread occurrence of the Barney animal, perhaps even coextensive with humans. Its certain identification may be complicated by morphological changes during its life cycle. It is possible that the development of the fuzzy epidermis, and the coelom separating it from the skeleton, are characters which form at sexual maturity. The juvenile stage may be exhibited solely by an immature hominid form, which presents very serious questions as to the correct identification of human children.

Notes

1. The order of authors was determined by lot.
2. The order of authors was determined without my knowledge.
3. The order of authors was determined by the last person who had the manuscript.
4. Academy of Natural Sciences, Philadelphia, Pennsylvania.
5. Freshwater Molluscan Research, Sewell, New Jersey, which of course has nothing to do with hominid evolution.
6. We refer to the genus Homo, but it reads less well to discuss "the rise of Homo"; to speak of "the rise of Man" is far more noble.
7. Virtually every issue since 1888.
8. Public Television Network. The program "Barney" depicts the animal in close, compatible relationship with children. Together they sing, dance, and discuss world issues of concern to pre-school age children. By logical induction, there is more than one specimen of this animal.
9. *The Dinosauria*, D. B. Weishampel, P. Dodson, and H. Osmólska (eds.), University of California Press, 1990.
10. That Barney could be a dinosaur is acceptable as an hypothesis, even though the dinosaurs are otherwise thought to be extinct. We initially believed that Barney could be descended from dinosaurian lineages, much as birds are thought to be so descended from these reptiles.

The Sad Crab
of South Africa

This graced the cover of *AIR* 1:6 (November/December 1995).

This photograph shows a sad-looking crab (*Ovalipes punctatus*) found on a South African beach. It was submitted by Michael Power, Department of Pediatrics & Child Health, Red Cross Memorial Children's Hospital, Rondebosch, South Africa.

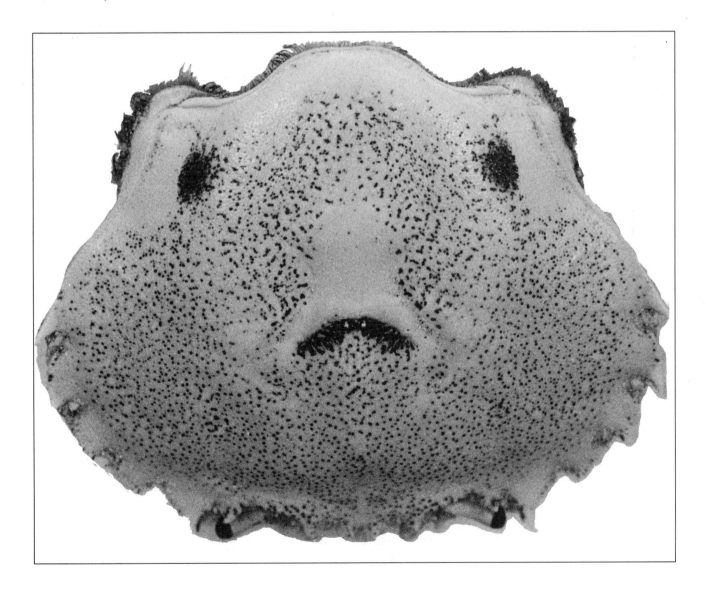

The Ability of Woodchucks to Chuck Cellulose Fibers

by P. A. Paskevich and T. B. Shea
Harvard Medical School
Boston, Massachusetts

This appeared in *AIR* 1:4 (July/August 1995).

How much wood could a woodchuck chuck if a woodchuck could chuck wood? We tackled this age-old question.

Marmota monax is a North American marmot, a ground-dwelling member of the squirrel family *Siuridae*, order *Rodentia*. *M. monax* is perhaps best-known by its nicknames: groundhog and woodchuck. The first is associated with an almost mythic ability to prognosticate seasonal weather variations. This study attempts to answer two questions associated with the latter: can a woodchuck, in fact, chuck wood, and , if so, can the chucked material be accurately quantified?

Materials and Methods

Twelve adult male *M. monax* were obtained through various means for this experiment. All were approximately 65 cm in length, with a 15 cm tail. On average, they weighed 6.5 kg. The experiment was conducted over a two-week period.

The Oxford Unabridged Dictionary of the English Language (3.2 kg) was used to define the word "chuck." Within the phrase "How much wood could a woodchuck chuck, if a woodchuck could chuck wood," we estimate the probability of the word "chuck" at 80% as referring to mastication and ingestion, and at 15% as referring simply to throwing wood around. We have somewhat arbitrarily assigned a 5% probability to the word referring to vomiting, and have designated the analysis of this particular component of the study to a first-year graduate student. (As an undergraduate this student majored in Business Administration and was Social Chairman of his fraternity; we therefore feel confident in qualifying him as an expert in the field.)

Each animal was housed in a 3.38 cubic cm cage, and deprived of all nourishment for seven days. At the end of this period, a 5.08 cm × 182.88 cm pine framing stud (herewith referred to as a 2 × 4) was fed through a pair of 5.08 cm × 10.16 cm hole in the sides of each cage at a constant rate (the Planks Constant) of 0.015 meters/hour for an additional seven days. Animals were videotaped over the entire fourteen-day period and rated by an independent observer as to whether they tried to: (a) eat the 2 × 4; (b) throw the 2 × 4 around the cage; or (c) throw up.

Results

One hundred percent of the experimental group tried to eat the 2 × 4 (Figure 1). We therefore conclude that "chuck," in the metaphorical sense, refers to eating, and further, that *M. monax* can, in fact, chuck wood. We also infer that several of the animals probably would have thrown the 2 × 4 at the independent observer if given a chance.

On average, *M. monax* mastication reduced 27% of the total stud volume to digested cellulose components. This amounted to 2533.465901 cubic cm per

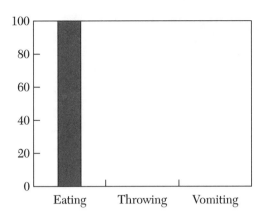

Figure 1: Percent of experimental group exhibiting "chucking" activities.

2×4 over the course of the experiment. The average chuck rate (ACR), therefore, was 361.9237001 cubic cm per animal per day.

As a control, fecal material from the twelve experimental animals was collected, examined and weighed. The ingestion/excrement ratio of 1.2 corresponds well to previously collected data ($p<.0005$).

Discussion

The question of ability and quantity (and/or rate) of wood chucking by *M. monax* is an issue that remained unresolved for decades. As we have shown, *M. monax* can in fact digest cellulose. Furthermore, the quantity and rate has been determined unequivocally.

This study, as is true of all experimentation subject to rigidly defined protocol, raises a number of intriguing questions that may serve as further avenues for exploration and funding. For example, it was noted that after fourteen days the woodchucks were noticeably less friendly to the independent observer. We postulate from this that a high-fiber diet may be causally related to personality changes; we have further determined that this trait can be transmitted interspecially (the graduate student also became noticeably more surly).

We also noted that by the conclusion of this experiment, all twelve woodchucks appeared noticeably thinner. Several had, in fact, lost weight. This may have been due to the enforced regimen of a single dietary food source, offered *ad libidum* but without variety. We have begun to explore the effect of variety in the woodchucks' diet. Our new experiment is identical to the first one, except that it substitutes a 60.96 cm × 121.92 cm × 1.905 cm sheet of plywood for the 2×4 framing stud.

Conclusion

Marmota monax is able to chuck wood at a rate of 361.9237001 cubic centimeters per day.

Acknowledgments

The authors wish to thank the first-year graduate student (whatever his name is) for collecting and examining the vomitus and fecal material, and for reviewing all 336 hours of videotape.

The *mickeymouse* Gene

This appeared in *AIR* 1:1 (January/February 1995).

This is an electrophoresis gel prepared from the so-called *mickeymouse* gene. According to its discoverers, who prepared the photograph, the gene encodes a novel cytosolic protein containing a triple repeat of the eight amino acid motif Methionine, Isoleucine-Cysteine-Lysine-Glutamate-Tyrosine-Methionine-Serine-Glutamate. Submitted by Timothy P. Angelotti and Marco A. Scarpetta of the University of Michigan Medical School, Ann Arbor, MI. Despite several requests, this photograph was not introduced as evidence in the O. J. Simpson trial.

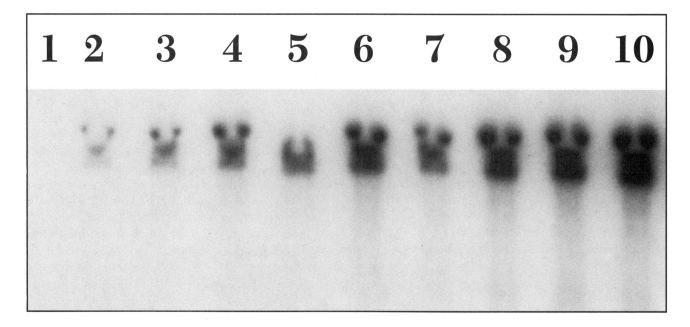

A Natural History of the Articulated Lorry, *Vehiculum articulatum* (Peterbilt, 1923)

by Angela E. Close
Department of Anthropology
University of Washington, Seattle, Washington

This appeared in *AIR* 1:2 (March/April 1995).

The articulated lorry is the largest insect species known to science. Although, in terms of sheer size, the articulated lorry could rightfully take its place in any assemblage of Pleistocene megafauna, the species seems not to have existed in its largest form before the earlier years of this century. Its evolution can be traced back through a remarkably complete series of primitive fossil ancestors to the third millennium BC in western Asia (Close 1982).

Physical Description

Extant forms of the articulated lorry are very large, routinely measuring over 10 m in length; the subspecies, *V. articulatum appendatum* or articulated lorry with trailer, may reach almost twice that size.

Articulated lorries normally have an elongated, rather rectilinear body. The segmentation between the head and thorax is always very pronounced (as is reflected in both the scientific and common names of the species), but that between the thorax and abdomen can be almost undetectable; indeed, the abdomen itself is frequently vestigial, probably a result of extreme selective pressures brought about by the articulated lorry's method of locomotion.

Locomotion

Although, like all insects, the articulated lorry has six legs, these take a form unique in the animal kingdom in that each is trifurcated and ends in a rod-like, horizontal foot. The juvenile articulated lorry builds up accretions of minerals and rubber into a disk-like covering around each foot. Given sufficient propulsion to overcome its tremendous inertia, the articulated lorry is then able to roll along the surface of the ground at speeds between 100 and 150 kph (Close 1988).

Under such stresses, the rubber element of the foot-disk may fail, providing scientists with vital data on the migration routes of articulated lorries. Some investigators have also tried to deduce these routes from the subfossil remains of the tire-striped armadillo, one of the most populous members of the Roadkill family (Viacaesidae); however, there is evidence that smaller members of the Vehiculum genus, such as *V. domesticum*, may occasionally prey upon the tire-striped armadillo and even upon the tread-impressed opossum.

With its highly developed means of terrestrial movement and its occasionally extreme weight, the articulated lorry, not unexpectedly, rarely takes flight and never does so deliberately. The wings have thus atrophied to small nubs at each side of the head, but retain their characteristic smooth and reflective surface.

Vision

The eyes are set in the front of the head, thus providing the stereoscopic vision essential for a creature given to moving at such speeds. Unusual among the insects, each eye has only a single facet. Among the most advanced subspecies, the two eye-facets approach each other very closely, leading earlier and misguided scholars to suggest that the articulated lorry might have but a single eye (Grayson 1970). Recent and more careful research has shown that, in all cases investigated, the two eyes remain physically separated by a structure known as the radio antenna.

Geographical Distribution

In a surprisingly short time, the modern form of articulated lorry has attained an almost global distribution, making it one of the most successful insect species to evolve within recent history. Since its spread has been so rapid, there has been little time for the processes of genetic drift to have their effect, and most articulated lorries, wherever on the globe they might be observed, are easily recognizable as members of the species, *Vehiculum articulatum.* Regional differences are, however, appearing, although they have yet to attain the force of specific differences. For example, in economically challenged regions throughout the world, the articulated lorry frequently sports a rich assemblage of flashing lights, lengths of tinsel and religious artifacts rarely seen on articulated lorries in northern Europe and North America. If these are related to mating patterns (Close 1992), then they may represent the first steps towards speciation.

Feeding Behavior

Although the articulated lorry frequently kills a variety of forms of Viacaesidae, there is little evidence that this is done for food, as has been claimed in the past (Grayson 1971). Indeed, the articulated lorry appears to be exclusively petrolivorous, and, despite the ubiquity of other potential sources, feeds only in particular places (in the vernacular, "truckstops"), along its migration routes. Here, articulated lorries gather in large numbers, frequently arranging themselves in long, straight, overnight lines. The reason for this behavior is little understood (but see below).

Reproduction

Almost nothing is known about reproduction among articulated lorries. Indeed, few scholars would claim to be able to determine the sex of an individual lorry with any certainty, although such claims have been made in the past (Grayson 1973; this should not be confused with his earlier [1969] and more popular paper on a related subject). Sexing of articulated lorries has traditionally consisted of attempting to lift the left, hind leg to see what might lie behind.

In light of the weight of articulated lorries, it is *ab initio* unlikely that mating would involve one animal mounting another. Mating therefore presumably takes place with the animals side-by-side. There have been no definite sightings of mating to confirm this, but some articulated lorries do spend considerable periods standing side-by-side for no other apparent purpose, and there is the further possibility that the overnight formations of articulated lorries seen at feeding areas might be some kind of group mating, distasteful though this is.

Migratory Patterns

The articulated lorry is a remarkably nomadic animal. Although it has been known for decades that the articulated lorry spends most of its life moving at excessive speed from one place to another, no overall pattern has ever been recognized. The most recent research on this matter suggests that this is because, except in feeding areas, the articulated lorry is basically a solitary animal and that migratory patterns should be sought at the level of the individual, since the "group" exists only in the absence of motion (Close 1993).

A similar approach might also do much to elucidate the mystery of why articulated lorries are frequently so heavy-laden. Such behavior is not otherwise unknown in the insect world, but even the dung beetle does not approach the level of transportational mania seen in the articulated lorry. A program of tagging and tracking individual animals has only just begun, but it is already apparent that each animal goes through alternating phases of being burdened and being unburdened. The reason for this redundant and uneconomic behavior is obscure, but may be related to reproductive fitness—that is, to sex.

Life Cycle

As one of the largest living land animals, the articulated lorry has no known predators. However, some of its behaviors appear to be defensive in nature, which suggests that predators might have existed in the past. These behaviors include the emission of excreta from blowholes at the back of the head, targeted so as to engulf a specific pursuer, and the imitation of a foghorn, with which it habitually routs smaller members of its own genus into nearby ditches. In the absence of predators, most articulated lorries presumably die of old age, probably in the third decade of life, since animals recognizably older than that are rarely seen. However, until scientists obtain the complete and unrecycled corpse of an articulated lorry for dissection, our knowledge of the biology and lifecycle of the species will remain in its infancy.

Acknowledgments

I wish to thank Professor Donald K. Grayson for first pointing out to me that articulated lorries are actually insects, an insight which has radically changed my view of the natural world. I have drawn heavily upon his pioneering work in the field, although most of it is wrong and all of it is out of date.

References

"Fossil insects of Mesopotamia: the earliest ancestors of *V. articulatum*." A. E. Close, *Journal of Sumerian Entomology*, vol. 23, 1982, pp. 56–67.

"The locomotion of the articulated lorry, *Vehiculum articulatum*, and the origin of speed limits," A. E. Close, *Journal of Insect Kinetics*, vol. 33, 1988, pp. 100–50.

"Migratory patterns among articulated lorries—there is no wood, only trees," A. E. Close, *Bulletin of Insect Metaphors*, vol. 46, 1993, pp. 65–8.

"Sexual variations in articulated lorries," D. K. Grayson, *SemiErotica*, vol. 2, no. 7, 1969, pp. 21–3 (with illustrations).

"Why Roadkill *(Viacaesidae)* die," D. K. Grayson, *Journal of Lorristics*, vol. 23, 1971, pp. 1–19.

"Expressions of sexual dimorphism in articulated lorries," D. K. Grayson, *Bulletin of the Society for the Study of Megafaunal Sex*, vol. 12, 1973, pp. 25–32.

Cyclic Variations in Grass Growth

by V. D. Irby, M. S. Irby

Department of Physics and Astronomy
University of Kentucky
Lexington, Kentucky

This appeared in *AIR* 1:4 (July/August 1995).

Grass exhibits a cyclical growth pattern surprisingly different from any other known plant. In this study, average grass blade heights have been measured, on a daily basis, over a 10 week period. Measurements were taken, utilizing vernier calipers, of the height of one hundred individual grass blades randomly chosen in a 10 foot square area positioned in front of an apartment complex in the Lexington, Kentucky area. (Measurements were also repeated with a different set of calipers to ensure reproducibility on a different apparatus). The average of these measurements was computed and experimental error was taken as the standard deviation of the mean divided by the square root of the number of grass blades used in the average. This procedure was repeated on a daily basis for a period of 10 weeks.

Results and Discussion

The average grass heights, measured in this work, are plotted as a function of time in Figure 1. As one can readily see, there exists a periodic variation in average grass height with an approximate cycle of 7 to 10 days. Another intriguing observation is that there exists a minimum grass height, or "grass baseline," of about 1.3 inches.

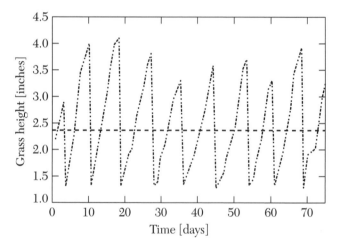

Figure 1: *Experimental measurements of average grass height are plotted versus time. Solid line represents experimental data. Short dashed line indicates a "constant grass height" calculation and is normalized to the experimental data to produce the best fit.*

Since the cyclic period of grass height is 7 to 10 days, one may conclude that grass height varies on a "week-about" basis. The physical mechanism responsible for this cyclic grass height phenomenon is not clearly understood at this time.

Arivederci, Aroma: An Analysis of the New DNA Cologne

by Jon Marks
Department of Anthropology
Yale University
New Haven, Connecticut

This appeared in *AIR* 1:2 (March/April 1995). A year later, the creator of DNA cologne was honored with the Ig Nobel Prize for chemistry.

It is widely believed that the most prevalent physiological form of DNA is a double helix.[1] Though you can't actually see it with your own eyes,[2] this has nevertheless proved to be a singularly valuable hypothesis in the study of DNA function and its manipulation in the laboratory.

For all the academic interest (and, more recently, venture-capital interest) in DNA, however, relatively few people have actually ever wanted to smell like it. The popularity of "Jurassic Park" seems to have changed all that, and to have given everyone's favorite biomolecule new life among the scent-buying public. Bijan Fragrances of Beverly Hills has responded to this demand, and has produced a new men's cologne called "DNA."

Its slogan is "Dare Feel Everything," though that would be abbreviated to DFE, which is not the name of this cologne. The connection is thus a bit unclear. There is no indication, for example, that the inspirational DNA is derived from neuronal cells.

The cologne does not smell very much like DNA, I am pleased to note, and in fact smells rather better than DNA. However, while researching this article I serendipitously discovered that "Poison" doesn't really smell like poison, nor does "Stetson" smell like a hat. There is no indication that DNA itself is used in the preparation of this cologne, or that the scent derives its name indirectly, as a concoction made from the bodily secretions of molecular genetics lab technicians or post-docs.

So, whence the name in this case? Has molecular genetics suddenly become such a turn-on that the mere mention of the hereditary molecule is expected to send the opposite sex into erotic paroxysms?

DNA. DNA. DNA.
Is it working?

Apparently, it is merely the bottle that actually merits the name. DNA is very strikingly suggested by the blue glass helix in which the scent is distributed. And a very attractive bottle it is (Figure 1). Though lacking

Figure 1: DNA cologne bottle.

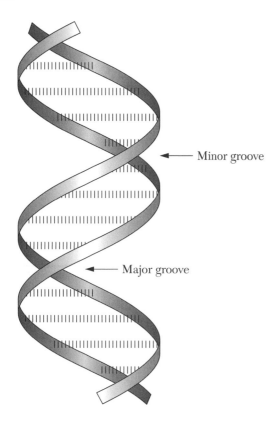

Figure 2: The common β-form of the DNA double helix, showing major and minor grooves. Stippled lines indicate Watson-Crick base-pairs.

Minor groove

Major groove

Naming a cologne "DNA" is presumably intended to be evocative of the power over nature possessed by the molecular genetics community. Wear the cologne, and you can exercise comparable power over the nature of the opposite sex. And as long as your partner hasn't taken molecular genetics, and doesn't study the bottle carefully, it could work. Otherwise, he or she might just start giggling.

We anxiously await the scent "*E. coli*," with certainty that its natural fragrance can be improved upon.

References

1. *Genes*, V. B. Lewin, Oxford University Press, New York, 1994.
2. Cricenti *et al.*, *Science* vol. 245, 1989, p. 1226.
3. J. D. Watson, and F. H. C. Crick, *Nature*, vol. 171, 1953, p. 737.
4. L. Pauling, and R. B. Corey, *Proceedings of the National Academy of Sciences*, vol. 39, 1953. p. 84.
5. Alternatively, the perfumers could have derived their design from the rare H-form of DNA (Mirkin *et al.*, *Nature*, vol. 330, 1987, p. 495; Rajagopal and Feigon, *Nature* , vol. 339, 1989, p. 637). This seems unlikely, however, as the triplex *H*-form of DNA does not appear to be well-enough characterized to base a cologne bottle on. Further, it requires Hoogsteen base-pair formation, which sounds awfully silly.

the structural grooves characteristic of the common or β-form of DNA, the bottle does retain the right-handedness of the helix, with strands oriented in the southwest-to-northeast direction (Figure 2).

Of considerable interest, however, is the *number* of strands of DNA used as a template for the bottle, revealed most clearly when examined end-on. Though, as noted above, most scientists have been working from the Watson-Crick model of DNA as a *double* helix,[3] Bijan's cologne bottle is actually a triple helix, consisting of three intertwined strands. This represents the apparent resurrection of a model proposed just prior to the Watson-Crick model, namely the three-stranded DNA model of Pauling and Corey[4,5] (Figure 3). Alongside the multi-decade lag between Mendel's experiments and the acceptance of those results, the three-stranded DNA model may have finally earned a place in the pantheon of "ideas before their time." It may have flopped among scientists, but four decades later, it is a hit with perfumers.

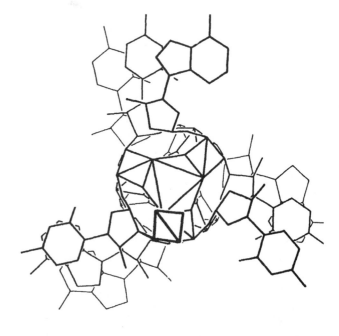

Figure 3: DNA model, after Pauling and Corey, 1953.

Happy Yeast

by Marcela A. Valderrama
and Jennifer Yang
Biology Department
Massachusetts Institute of Technology

This appeared in *AIR* 2:1(January/February 1996).

We announce the discovery of a new strain of *S. cerevisiae*, a phenotypically "happy" yeast.

The phenotype was expressed after plating *S. cerevisiae* transformants onto media that was supposed to be synthetic-complete-urea. The actual content of the media is not known.

No colonies grew on the plates, but the manner of spreading yielded this cluster of cells.

This supports our theory that biology is the happiest science.

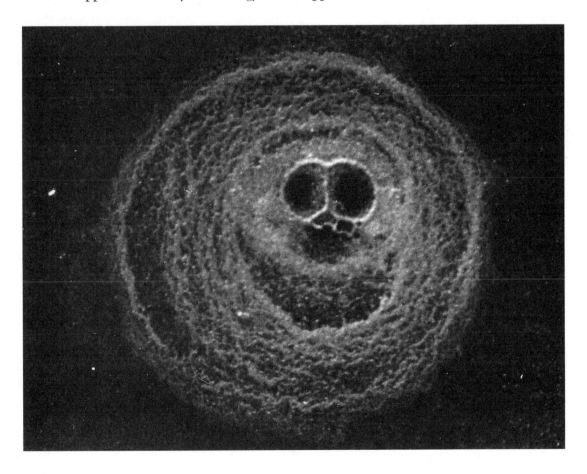

A Man, a Woman, a Yeast

by Alice Shirrell Kaswell and Arsenio Pfist
AIR staff

This appeared in *AIR* 1:5 (September/October 1995).

Saccharomyces cerevisiae's activities are of almost infinite variety. Perhaps nowhere is this better illustrated than in the research paper, "Bread-Making as a Source of Vaginal Infection with *Saccharomyces cerevisiae*: Report of a Case in a Woman and Apparent Transmission to Her Partner," J. D. Wilson, B. M. Jones and G. R. Kinghorn, *Journal of Sexually Transmitted Diseases*, Jan-Mar 1988, pp. 35–6. The abstract reads:

We report a case of a woman, who baked her own bread, acquiring a vulvovaginal infection caused by *S. cerevisiae*, a yeast used extensively in the baking and brewing industries. Her partner was also infected and the infections proved difficult to eradicate. Oral and topical treatment with mystatin along with disinfection of both patients' underwear produced clinical and microbiologic cure.

Nematodes and Hieroglyphs

by Mark Benecke

Zoologisches Institut der Universitaet
Koln, Germany

This appeared in *AIR* 1:2 (March/April 1995).

Hieroglyphs were not invented by the Egyptians. Credit must, instead, be given to the microscopic roundworm *Caenorhabditis elegans*. Today, this modest creature is best known for its role in the study of developmental biology and genetics.[1,2] In the laboratory, *C. elegans* displays a wide range of behaviors. My research shows that the range is much wider than we had realized.

Materials and Methods

Worms were bred in dishes filled with bacteria-covered agar. A hermaphrodite about 1/2 mm in length, *C. elegans* fed on a layer of *Escherichia coli*. The worms were placed in a phosphate buffer. Two gentle centrifugation steps were then taken to concentrate the animals.[3] Worms were then transferred to a polylysine-coated slide and frozen down to −210°C in liquid nitrogen. After thawing, the roundworms were analyzed optically using Nomarski microscopy.

Results

The worms formed their bodies into distinct shapes. Their postures depended on the ambient breeding conditions.

The most interesting pattern was formed after a prolonged shortage of food. Groups of three worms met to jointly form the hieroglyph meaning "eternity" (Figure 1A). At the same time, other roundworms started to form the hieroglyphic symbol calling for one to "experience" (Figure 2B). The combined meaning is unmistakable: "experience eternity." Figures 1B and 2B show the corresponding hieroglyphs.

A

B

*Figure 1: Egyptian hieroglyphs are stylized versions of the shapes formed by the microscopic roundworm Caenorhabditis elegans. (**A**) Here, three worms join together to form the shape meaning "eternity." (**B**) This photo shows the corresponding hieroglyph.*

A

B

*Figure 2: (**A**) A starved specimen of Caenorhabditis elegans is seen forming the shape that means "experience." (**B**) This photo shows the corresponding hieroglyph.*

Under favorable breeding conditions, the worms formed a very different set of shapes similar to the hieroglyphs meaning "govern," "boast," and "beat the drum" (figures not shown).

Discussion

Soil roundworms, such as *Caenorhabditis elegans*, are common all over the world and have been so for more than five hundred million years. Under poor breeding conditions, the worms form hieroglyph motifs alluding to death and transitory events. Under good conditions, the worms express positive—sometimes enthusiastic—motifs.

As we know from Bertelsmann,[4] it was the Laryngaeans who handed down the hieroglyphs to the ancient Egyptians. The people of Laryngaea almost certainly were able to read the roundworm messages with the unaided eye. Aristotle indirectly[5] vouched for this when he wrote that "the Laryngaeans are the most sharp-eyed people on earth."

Taking these facts together, it seems likely that the first printed characters were developed by free-living roundworms. Furthermore, it is likely that Champollion made use of *C. elegans* when he deciphered the Rosetta Stone in 1822,[6] simply comparing the hieroglyphs with roundworm shapes that he generated under well-defined, well-controlled conditions.

Finally, we would like to draw the reader's attention to the complete hieroglyph text formed by starving nematodes. They ask the observer to "experience eternity," a proposal which has a certain charm, and which was followed by most people living in ancient times.[7,8]

References

1. "A worm at the heart of the genome project," R. Lewin, *New Scientist*, vol. 25, no. 8, pp. 38–42.
2. "Der Nematode *Caenorhabditis elegans*: Ein entwicklungsbiologischer Modellorganismus," E. Schierenberg and R. Cassada, *Biologie in unserer Zeit*, vol. 16, no. 1, 1986, pp. 1–7.
3. "The genetics of *Caenorhabditis elegans*," S. Brenner, *Genetics*, vol. 77, 1974, pp. 71–94.
4. "Volkslexikon," Lexikon-Redaktion Bertelsmann, *Hieratische Schrift*, Bertelsmann, Guetersloh, 1956.
5. *Libelli, qui parva naturalia vulgo appellantur*, Aristotle, Brummenius, Paris, 1560.
6. *Principes généraux de l'écriture sacrée égyptienne appliquée á la représentation de la langue parlée*, J. F. Champollion, Institut d'Orient, Paris, 1841.
7. *Wer ermordete Mozart? Wer enthauptete Haydn?* E.W. Heine, Diogenes, Zurich, 1986.
8. *Luthers Floh*, E. W. Heine, Diogenes, Zurich, 1990.

The Surfer Girl Fungus

This graced the cover of *AIR* 2:2 (March/April 1996), which was the annual Swimsuit Issue.

This photomicrograph shows spores of the fungus *Sporothrix flocculosa*. The fungus was grown on a sterilized human hair. The spores fell and were dehydrated, forming the pattern shown here. Submitted by Don Pohlman of Windsor, Ontario.

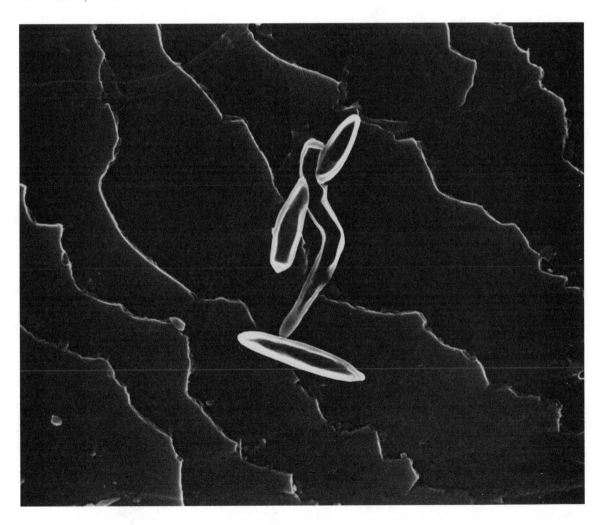

Scientific Dining

Blackford Hall
Cold Spring Harbor Laboratories
Cold Spring Harbor, New York

by **Karen Hopkin**
National Public Radio
Biochemist and food critic

This appeared in *AIR* 1:3 (May/June 1995). It was the first of an ongoing series of reviews of cafeterias at the world's great research institutions.

The decor is stoic but pleasant in a dining hall that offers a spectacular view of the quaint and peaceful Cold Spring Harbor. With its informal ambiance, Blackford Hall draws a very faithful following. In fact, many diners return just about every day.

"Actually, the food here is not bad, really," raves geneticist Alcino Silva, a frequent diner. "At least they don't rip you off. . . . They offer mediocre food at a mediocre price."

The group we lunched with began the meal with a dish called "Shrimp Nuremberg." This entree was described by the diners as being "chunky," "yellowish," and "somewhat recognizable," with a taste that was "subtle, sort of."

The weekly menu frequently features ethnic dishes, ranging from jambalaya to lamb curry, and lyonnaise potatoes to white beans and sausage with corn chowder. We were told that these meals usually proved to be less frightening than predicted.

We were most pleasantly surprised by the dessert selection. The cakes and pies, imported from a local bakery, were described as "supreme" and "highly recommended," though when it came to dessert, resident scientists seemed to feel that quantity was

Figure 1: *Blackford Hall is the social and fine dining center of Cold Spring Harbor Laboratory on scenic Long Island.*

as important as quality. Because of the imprecise nature of the cake-cutting procedure, biochemist Yuri Lazebnik informed us that, with careful observation and selection, one could choose a slice of cake that might be two standard deviations larger than the average hunk. All for the same price, of course.

Unfortunately, we could not stay long enough to experience firsthand the boisterous excesses of the legendary Saturday night lobster banquet. Or the warm comfort of the Sunday afternoon lobster bisque. Or the half-price bargain of Monday's lobster salad.

The quality of the food improved exponentially after Chef Ron Padden, formerly of the Pierre Hotel in Manhattan, joined the staff in March of 1994. He replaced a chef who had been the head cook on a submarine for seven years. "He certainly had a captive clientele," said geneticist Michael Hengartner of their former chef, "But he wasn't too good with fresh fruit."

Hengartner summed up the Blackford experience most eloquently. "It's the best place for miles around," he said. "Actually, it's the only place for miles around."

Figure 2: The cafeteria's internal design and layout, seen here in the 1920s, has worked so well that it remains virtually unchanged in the 1990s. Both photos courtesy of the Cold Spring Harbor Laboratory Archives.

Ratings Quality: 1.78 Trendiness: 2.5 Bearded Men: 3

Explanation of ratings:

Quality: Food quality is rated on a scale from i (the square root of -1) to pi (with a numerical value of 3.141592 …), where a rating of i signifies that the food is of good quality only in your imagination, and pi signifies that it is roundly accepted as being delicious. As an example, the research facility with the finest quality food, a *pi* rating, would be the Howard Hughes Medical Institute headquarters in Chevy Chase, Maryland; the cafeteria with the poorest quality food, a rating of i, would be the Princess Margaret Dining Hall at the University of Swansea in Wales.

Trendiness: The cafeteria's trendiness is also rated on a scale from i to pi. As an example, the dining hall where one would most like to be seen would be at the Karolinska Institute in Stockholm, Sweden. The cafeteria where one would least like to be seen would be the Princess Margaret Dining Hall at the University of Swansea in Wales.

Bearded Men Index: Number of photos or drawings of bearded men displayed on the walls of the cafeteria.

Scientific Gossip

Contains 100% gossip from concentrate

compiled by Stephen Drew
AIR staff

These results are collected from various issues of *AIR*.

Hard to Swallow

A plump belly can have its advantages. **Christer Brönmark** and **Jeffrey Miner** of the University of Lund found that European carp develop enlarged midsections after several months of living in proximity to predatory pike. The carp grow wide enough that they no longer fit into their predators' mouths. Several studies are now being mounted to see whether and how this applies to humans. The prospective studies are unusual in that they will be funded by a consortium of German beer companies. (For pertinent information, see *Science*, November 20, 1993)

Food Fight in South America

Genetically engineered crops continue to inspire emotional debate about food safety, agricultural efficiency, and how to tackle the problem of world hunger. In an effort to end the skirmishing between various groups, **Dr. Robert Richard Smith** and several colleagues attended last month's Buenos Aires meeting on the subject. Smith is founder of Non-Extremists for Moderate Change (NEMC), an organization whose goal is expressed in its name. In every country except Finland, NEMC members have been greeted with scorn, projectiles, violent attack, and police arrest as they attempt to find solutions to various intractable problems. At the Buenos Aires meet-

ing, food purity activists, bioengineering researchers, organic farmers, and public health experts all heaved bags of genetically engineered vegetables at the NEMC members before physically ejecting them from the meeting site.

Improper Plant

Linguistics purists are mounting a campaign to restore general use of the name rapeseed over its newer, commercially imposed moniker, canola. Formally known as *Brassica napus*, the plant has become popular because its oil is rich in the monounsaturated fats favored for healthful cooking. However, marketers believe that the popularity of rapeseed oil over olive oil and other alternatives owes more to the change of name they imposed some years ago than to the oil's biochemical properties.

Cruel Milk

Most cats are to some degree lactose intolerant (i.e., unable to digest milk sugar). The British Animal Ascendency Society has now stated that owners who feed milk to their pets are committing a chronic form of cruelty and should threfore be put to death. The group, founded in 1987 to protest animal experimentation, has been expanding the scope of its activities. Last month the Society's director, ex-supermodel

Brent Stith, called for the imprisonment of "any animal, human or otherwise," that "engages in carnivorous activity."

Testing the Limits to Growth

With earth's population growing by approximately 100 million people per year, some economists argue that "for every new mouth to feed there's a new pair of hands to work out a means of feeding them." A new experiment will test the theory. Modeled after the Biosphere experiment, the PopuSphere will be a sealed, Plexiglas-enclosed structure located in the Arizona desert. The trial period will begin with a near-theoretical-capacity of 200 individuals, all of prime child-bearing age and possessing firm religious beliefs against contraception. Experimenters will seal the subjects into the PopuSphere next summer, then measure the population twenty years later.

May We Recommend

Items that merit a trip to the library

compiled by Stephen Drew
AIR staff

These items are collected from various issues of *AIR*.

Darkness Under Light

"Use of Floor Polish in Mounting Sections for Light Microscopy," K. M Imel, *Microscopy Today*, issue #95-1, 1995. *(Thanks to Gail Celio for bringing this to our attention.)* The report reads in part:

> Over the years a number of methods have been described to mount tissue sections on glass slides. This paper presents a novel mounting medium that is inexpensive and readily available at the nearest grocery store—floor polish. The floor polish has demonstrated a much lesser degree of tissue distortion in some tissues as compared to the mounting media previously used in our laboratory, including Permount® (Fisher Chemical, Fair Lawn, NJ) and 30% sucrose in distilled H_2O.

Deep Fuzz

"Implications of Lint Deposits in Caves," Pat Jablonsky, *NSS News*, vol. 50, no. 4, April 1992, pp. 99–100.

The Origin of Asses

"Oldest ass recovered from Olduvai gorge, Tanzania, and the origin of asses," C. S. Churcher, *Journal of Paleontology*, v. 56, no. 5, 1982, pp. 1124–5. *(Thanks to Andrew MacRae for bringing this to our attention.)*

Muttering Mutton

"The perception of speech sounds recorded within the uterus of a pregnant sheep," Scott K. Griffiths, W. S. Brown, Jr., Kenneth J. Gerhardt; Robert M. Abrams, and Richard J. Morris, *Journal of the Acoustical Society of America*, vol. 96, no. 4, Oct. 1994. *(Thanks to Lucy Horwitz for bringing this to our attention.)*

Inefficient Beef Roast

"Foreign body caused endometritis in a cow," Von A. Boos and D. Ahlers, *Dtsch. tieraztl. Wschr*, Sept. 1994, pp. 341–80. *(Thanks to R. S. Youngquist for bringing this to our attention.)* The authors report that:

> Uterus, uterine tubes, and ovaries of a cow were obtained from the local abattoir. . . . Post-mortem findings indicate a subacute to chronic endometeriosis, caused by a cigarette-lighter found in utero.

Milk and Hair

"The relationship between facial hair whorls and milking parlor side preferences," M. Tanner, et al. Abstract #797 from the *Journal of Dairy Science*, vol. 77, Annual Meeting Abstracts, 1994. *(Thanks to C. Robert Campbell for bringing this to our attention.)*

One Giant Leap for Frog-kind

"Behavior of Japanese tree frog under microgravity," A. Izumi-Kurotani, M. Yamashita, Y. Kawasaki, T. Kurotani, Y. Mogami, M. Okuno, T. Akiyama, A. Oketa, A. Shiraishi, and K. Ueda, *Biological Sciences in Space*, vol. 5, 1991, pp. 185–9.

Queasy Quasi-aquatics

"Motion sickness in amphibians," Richard J. Wassersug, Akemi Izumi-Kurotani, Masamichi Yamashita, and Tomio Naitoh, *Behavioral and Neural Biology*, vol. 60, 1993, pp. 42–51. The abstract reads in part:

> We explored the question of whether amphibians get motion sickness by exposing anurans (frogs) and-urodeles (salamanders) to the provocative stimulus of

parabolic aircraft flight. Animals were fed before flight, and the presence of vomitus in their containers after flight was used to indicate motion-induced emesis.

Unpleasantness

"Hypercapnia and hypoxia which develop during retching participate in the transition from retching to expulsion in dogs," Hiroyuki Fukuda and Tomoshige Koga, *Neuroscience Research*, vol. 17, 1993, pp. 205–15. *(Thanks to David Langleben for bringing this to our attention.)*

We welcome your suggestions for this column. Please enclose the full citation (no abbreviations!) and a photocopy of the paper.

AIR Vents

Exhalations from our readers

Note: These letters are collected from various issues of *AIR*. The opinions expressed here represent the opinions of the authors and do not necessarily represent the opinions of those who hold other opinions.

For the Children's Sake

I read Theriot *et al.*'s report on "The Taxonomy of Barney" with keen interest. There was some dinosaur research published in the last year or so about T Rex testicles (really) and how scientists think they retracted into the T Rex's body when not needed, which, given the general appearance of the beast, was probably pretty often. Upon reading this, I immediately thought of Barney, and realized what this would mean for the impressionable children of America if the lovable but pantless host were ever to become unexpectedly aroused. I would think this was a very important point to include in your Barney taxonomy research.

Nathan Bos
University of Michigan School of Education
Ann Arbor, MI

Difficult Delivery

I was surprised that Close ("The Natural History of the Articulated Lorry") indicated so little knowledge of reproduction in the lorry. It has been reliably observed (by me) that lorries BACK UP to each other and engage their external excretory tubules (a.k.a. tailpipes) in ways not dissimilar to those of many vertebrates. The young are carried internally by the female. Once born, the young congregate in long lines in specified nursing areas for each subspecies (e.g. *V. articulatum hondaensis*, *V. a. mercedes*). It is the metamorphosis from juvenile to adult form that is most puzzling, since the young do not show the segmentation typical of adults, as the head is continuous with thorax and abdomen.

Connie Sancetta
Washington, DC

All for One

Cavalli-Sforza (the well-known population geneticist) and Bowcock recently estimated[1] that "Europeans emerged only about 30,000 years ago, and appear to have 65% Asian, and 35% African ancestry (plus or minus 8%)." Since hearing this, I am gleefully using it on forms which I have to fill out (e.g., for Federal grants). On being admitted to hospital recently, it immediately elicited sympathy from the (charming, black) receptionist who put it all into the computer banks.

Leonard Genghis Khan Umslopogas Finegold
Physics Department, Drexel University
Philadelphia, PA

P.S. Does this explain why I like Indian and Chinese food?

1. The *New York Times*, July 27, 1993, p. C9.

Candy-Coated Viewpoint

A friend of mine is a neuro-anatomist doing extensive research in primate brain evolution. In the past he has utilized rhesus monkeys for this research and I'm here to tell you that when he got finished with them they were rhesus pieces.

Stephanie Vardavas
Washington, DC

CHAPTER 7
Medicine and Biology

Here are some meaningless questions. The fact that some people are obsessed by them is curious, very curious.

1. Why *do* many doctors insist that the letters "M.D." appear with their names in official published papers?
2. Why *don't* many Ph.D.s insist that the letters "Ph.D." appear with their names in official published papers?
3a. Is there a pecking order among: (i) medical doctors; (ii) research scientists; (iii) people who have *both* an M.D. and a Ph.D.?
3b. Can you make an enemy for life by calling certain people by the "wrong" titles?

This chapter presents some useful medical findings from laboratories, clinics, and detention centers.

Standards are important. For one doctor's diagnosis to be useful to another doctor, both people must agree on what they are talking about. Mike Dubik and Brian Wood's paper "How Dead is a Doornail?" establishes a firm—one might say stiff—basis for a standard that had previously been just a metaphor. Thomas Michel's "Politically Correct Guide to Cardiology" revises an entire group of old standards, making it more palatable to modern sensibilities. Michel's work also perhaps implies that one can discriminate between sense and sensibility.

Pheromones are scent chemicals that various living creatures use to convey signals to each other. The general public has just recently become aware of their existence and their power. Our "Mystery Pheromone Coupon" will be of use to any mammal who is still breathing.

Alexandru Stan's "A *glioblastoma multiforme* that Resembles Little Bird" and Robert Roger Lebel's "Fetal Man in the Moon" demonstrate how medical imaging can produce unexpected discoveries. There are many ways to look at a single object. Two score and ten of them are presented in Jeffrey Moran's "Fifty Ways to Love Your Liver."

In pediatrics, the human touch is extremely important. If you have children, we recommend that you bring a copy of G. L. Hansen's "The Medical Effects of Kissing a Boo-Boo" to your doctor.

All of us are concerned with the frightening rise in health care costs. *AIR* has several initiatives to keep those costs down. One of the simplest is presented in "The Pop-Up Medical Thermometer."

Everyone makes jokes about dentists. We do, too. For examples, see A. J. Tuversen and Stanley Rudin's "The Tomb of the Unknown Dentist" and Walter Martin's "The Dental Micro-Luger."

In many of this book's other chapters, the "May We Recommend" section presents helpings from the world's vast "serious" research literature. This chapter contains two such sections, "May We Recommend" and "Boys Will Be Boys." The latter became a regular column after we noticed how very, very, very many doctors were sending us clippings of medical reports that appeal to top-grade adolescent minds. Rather than change the tone of the rest of *AIR*, we opted to isolate the best of that material in a separate column. It is a matter of pride that the items we receive for the "Boys Will Be Boys" column come almost as often from women as from men. They deal with "research by and for adolescent males of all ages and sexes."

How Dead Is a Doornail?

by Mike Dubik, M.D.
Brian Wood, M.D.

This appeared in *AIR* 1:6 (November/December 1995).

For hundreds, if not thousands, of years it has been accepted as an axiom that inanimate objects, such as nails, are dead. This self-evident truth has been expressed in the phrase: "dead as a doornail." Thus, someone who is unequivocally dead is said to be "dead as a doornail."

Advanced life support technology now allows us to maintain the heart and lung's functionality in patients who no longer have any brain function. This ability has created legal, moral and religious conundrums. Until a generation ago, these problems were solely the domain of a few ethicists who entertained them as theoretical exercises.

However, now most states have laws concerning brain death. The American Medical Association, the American Bar Association, the American Neurological Association, and the American Academy of Pediatrics came together and formed a Special Task Force[1,2,3,4] and have endorsed the following as a definition of death: Irreversible cessation of all function of the entire brain, including the brainstem.

If the definition of death as expressed by the AMA *et al.* has validity, it should be possible to compare this recent criteria against the widely accepted and time-tested "doornail" standard. We did just that.

We subjected a large doornail (see Figure 1) that was forged in 1986[5] to thorough examination, prolonged close observation, and an electroencephalogram (EEG).

Our Findings

The doornail was repeatedly examined and closely observed over a 24 hour period.

Figure 1: The subject of the study.

1. The nail did not exhibit any vocalizations of volitional activity.

2. The nail evidenced no spontaneous eye movements; neither could respiratory movements be detected.

3. There was no evidence of postural activity (decerebrate or decorticate).

4. The nail made no spontaneous or induced movements whatsoever.

Thus, the nail met the "physical examination" criteria of death.[3,4]

A well-executed and reliably read electroencephalogram is a useful adjunct in the diagnosis of brain death. We performed a 30-minute EEG to document electrocerebral silence (see Figure 2). As is often the case with small children, it was not possible to meet the standard requirement for 10 cm

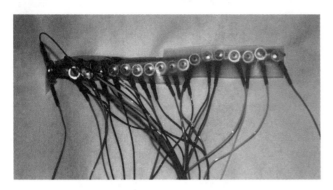

Figure 2: The doornail undergoing electro-encephalography.

electrode separation. Instead, the inter-electrode distance was decreased proportionally to the size of the nail's head. The EEG was isoelectric, i.e. flat. Further, there was no electrical response to rousing stimuli. When we subjected the doornail to rousing stimuli, there was no response.

We conclude that the criteria for death as described in modern medical literature[1,2,3,4] is valid and may be used with confidence by clinicians.

References

1. "Determination of brain death," Ad Hoc Committee on Brain Death (The Children's Hospital, Boston, MA), *Journal of Pediatrics*, vol. 110, January, 1987, pp. 15–19.
2. "Guidelines for the determination of death," President's Commission for the Study of Ethical Problems in Medicine and Biomedical and Behavioral Research, Washington, DC, *Journal of the Amercican Medical Association*, vol. 246, 1981, p. 2184.
3. Report of a Special Task Force: Guidelines for the Determination of Brain Death in Children, *Pediatrics*, 1987, vol. 8, no. 2, pp. 298–300.
4. "Guidelines for the Determination of Brain Death in Children," Task Force for the Determination of Brain Death in Children, *Neurology*, vol. 37, June, 1987, pp. 1077–8.
5. The nail was seven years old when the study was conducted.

This appeared in *AIR* 1:4 (July/August 1995).

Annals of Improbable Research (AIR)

Mystery Pheromone Coupon

This piece of paper makes you very attractive to others. It may contain chemical attractants, called pheromones, that make you irresistible to the opposite sex.

Fold it up and put it in your pocket. Keep it with you at all times.

WARNINGS

1. There is a fifty percent (50%) chance that this particular coupon will affect your own sex rather than the opposite sex.

2. May be flammable when combusted.

3. Do not take internally.

4. Do not take into airplanes or other confined spaces.

5. Do not remove from airplanes or other confined spaces.

6. Do not mail, as recipient's sex, and thus recipient's reaction to the coupon, is unpredictable.

7. Do not use when in the presence of insects of the opposite sex.

8. The manufacture of this product involved no cruelty to animals. However, its use in the presence of mammals—especially bovines—cannot be guaranteed to be cruelty free.

9. We are not responsible, etc.

10. Have a nice day.

A *glioblastoma multiforme* that Resembles Little Bird

This graced the cover of *AIR* 1:1 (January/February 1995).

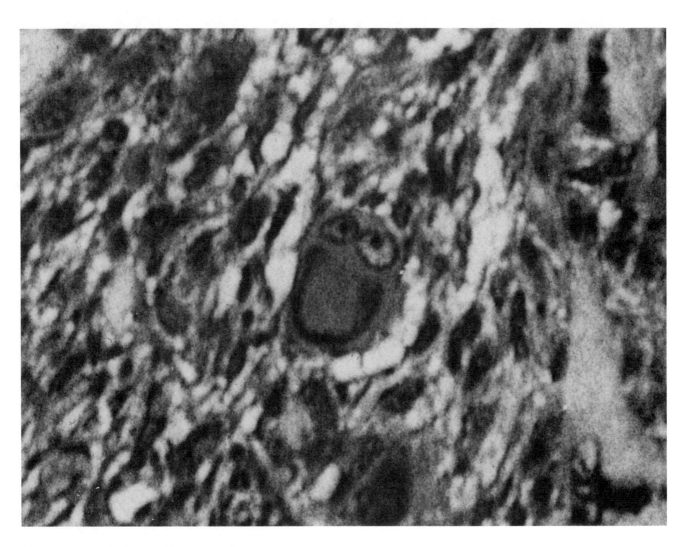

Photomicrograph of a Glioblastoma multiforme.
Submitted by Alexandru C. Stan, M.D. Medical
School, Hannover, Hannover, Germany.

Fifty Ways to Love Your Liver

by Jeffrey B. Moran
(with apologies to Paul Simon)

This appeared in *AIR* 1:4 (July/August 1995).

This organ's not inside your head,
she said to me.
It's in the abdomen, anatomically.
I'd like to help you understand hepatically,
There must be fifty ways to love your liver.

She said, it's really not where food
that has been chewed
Is digested into tiny molecules.
But it makes bile which into the gut is spewed.
There must be fifty ways to love your liver.

Chorus:
You just lay off the smack, Jack.
Eat some more bran, Stan.
Make the right choice, Royce,
Just listen to me.

Cut out the brew, Sue.
Don't want to be yellow, fellow.
Eat your protein, Gene,
And let your liver be.

She said, hepatitis pains your liver so,
It usually comes from viruses, you know,
In contaminated food and used hypos.
There must be fifty ways to love your liver.

She said, remember how cirrhosis
makes you bawl.
It comes from drinking too much alcohol.
And I realized that though she had a lot of gall,
There must be fifty ways to love your liver.

Chorus: as before

The Medical Effects of Kissing Boo-Boos

by G. L. Hansen
Department of Parapathology
Minnesota State University
Minneapolis, MN

This appeared in *AIR* 1:5 (September/October 1995).

It is common practice among parents, when a child receives a mild injury, to apply mild topical remedies followed by activities and verbal statements intended to reassure the child. Frequently, the parent will apply osculatory pressure (i.e. a kiss) to "make it better." Although this form of treatment is ancient in origin, and is no doubt thought to be efficacious by those who practice it, there has been considerable doubt among clinicians as to whether kissing a wound as an aide to the body's natural healing processes is a medically effective course of action.[1,2]

A number of pertinent questions present themselves. Is it necessary for the child's mother to administer the treatment, or is a father or other blood relative sufficient? Does the specific verbal reassurance have any effect? If so, how?

The Osculation Study

To test the effectiveness of various forms of osculatory wound treatment, the following experimental protocol was used.

For a period of 18 months, we monitored areas where pediatric wounds were likely to be found. In practice, this was usually on playgrounds near swings and slides, under tree houses, and at the bottom of steep hills where children were likely to ride bicycles. If a child suffered an injury, a researcher followed him or her home, and after securing a form of consent from the parent, observed and recorded the treatment process.

A patient (accompanied by the caregiver) reports for a follow-up examination one week after receiving osculation treatment. Photo: Bruce Petschek.

Of the 24,617 candidate wounds in the study, consent was obtained for 23 cases. About half of the parents applied osculatory treatment, while half did not. This fact, along with other relevant circumstances, was noted at the time of treatment. Subsequently, daily follow-up interviews were conducted until the healing process was deemed to be complete.

Basic Results

The data show unambiguously that kissing a wound does indeed make it better. For all types of wounds, application of osculatory pressure at the time of treatment shows an average 5.2 day reduction in the average healing time of minor pediatric wounds.

The beneficial effects of osculatory therapy are clear. The extremely slow healing seen in the last category in Table 1 is remarkable; it remains an un-explained anomaly.

Table 1. Wound Categories

Type	Severity	Quantity
Type A ("Booboos")	1	12
Type B ("Ouches")	2	7
Type C ("Bad ones")	3	3
Type F ("Faking")	?	1

Table 2. Healing Rates By Wound Type

	Healing time (days)	
Type	With Osculation	Sans Osculation
Type A ("Booboos")	21.7	26.9
Type B ("Ouches")	24.5	29.7
Type C ("Bad ones")	25.7	30.9
Type F ("Faking")	56.0	–

The Caregiver

In some cases, the treatment was administered not by the child's mother, but by another adult caregiver. The data show that under most circumstances, the relationship of the person giving the treatment to the child is immaterial.

Table 3. Healing Rates By Caregiver

Caregiver	Quantity	Healing time (days)
Mother	8	26 ± .6
Father	9	26.3 ± .8
Aunt	5	27.2 ± .6
Nanny	4	26.5 ± .5
Older brother	1	56.0 ± 0

The Dressing

We found that the type of dressing used had little correlation with healing rate. Of the different types of dressings used, most were indistinguishable from this point of view.

Table 4. Healing Rates By Dressing Type

Dressing	Quantity	Healing time (days)
Johnson & Johnson Band-Aid (Sheer)	6	26.3 ± .7
Johnson & Johnson Band-Aid (Flesh)	7	26.5 ± .8
Curity Ouchless	9	27.1 ± .5
Smurf Bandage	1	56.0 ± 0

Summary and Recommendations

In most cases, kissing a wound does indeed "make it all better." This is true for most wounds of differing severity. The effect is largely independent of who administers the treatment.

In most cases, the type of dressing used made little difference, however, certain dressing types had healing rates much worse than the norm. It is recommended that clinicians avoid the use of "Smurf" bandages in critical situations.

The benefits of kissing a child's wound are, perhaps, somewhat surprising given the lack of any identifiable physiological mechanism. However, we have no quarrel with this age-old practice, and find it to be clinically efficacious. This, at least, is a comfort.[3]

References

1. "Osculation revisited," Melvyn Sneath, *Maledictus,* vol. 47, June 1987, p. 3523.
2. "Excess Mortality Due to Pendiculation," Otto Grompus, *Death and Morbidity Quarterly,* January 1991, p. 23.
3. "Points to Ponder," *Reader's Digest,* July 1990, p. 153.

Ig Nobelliana

Words for the ages

"I wish my mother could be here to see me accept this prize."

—Dr. James Nolan, co-winner of the 1993 Ig Nobel Medicine Prize for his painstaking medical report, *Acute Management of the Zipper-Entrapped Penis*

Fetal Man in the Moon

This appeared in *AIR* 1:2 (March/April 1995).

This ultrasonographic image was obtained during an obstetrical transvaginal study at five weeks of gestation. The black area toward the left is the fluid-filled gestational sac. The small white circle inside the fluid region is the yolk sac. The white area toward the right is the trophoblastic reaction coalescing to a placenta. The embryo is not visible here, but was seen in other views to be developing normally. Photo courtesy of Robert Roger Lebel, Genetics Services, Elmhurst, Illinois.

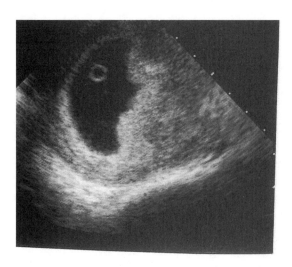

The Guide to Politically Correct Cardiology

by Thomas Michel, M.D., Ph.D.

Cardiology Division
Brigham and Women's Hospital
Harvard Medical School
Boston, Massachusetts

This appeared in 1993.

The politically correct cardiologist should become aware of the evolving terminology in the field, as it seems likely that in the near future physicians will not be permitted to bill for charges incurred by patients diagnosed with politically incorrect illnesses.

It has been widely acknowledged that the nomenclature of the specific diagnoses applied to a particular patient's ailment may profoundly affect the patient's well-being, self-image, and time to full recovery. It seems only logical that diagnoses should be empowering rather than belittling. They should reflect the patients' ability to transcend their illness rather than be dominated by the judgmental, impersonal and paternalistic semiotics of the medical profession. The nomenclature of heart disease is particularly troubling in this context.

Table 1 summarizes a variety of commonly applied cardiac diagnoses, and supplies a translation into politically correct terminology.

Heart failure is probably the most common cardiac diagnosis, but the very term is belittling. It says to the patient, "your heart has *failed*." It is far more empowering to term the patient "inotropically challenged" to reflect the fact that the contractile (inotropic) state of the heart shows room for improvement. Likewise for diastolic failure, the fact that the heart doesn't relax (lusotropy) as well as it might should not be branded as "failure."

Patients have long been termed as having "sick sinus syndrome" if the principal pacemaker (sinus node) cells of the heart beat too slowly. This illness sounds rather more like an upper respiratory infection, and applies the value judgment of *sickness* to a noble, organic cardiac structure. Better to say that a patient is "chronotropically challenged" or "systolically impaired."

The key here is that cardiac diversity is to be celebrated, not denigrated. The term "aberrant" used to be applied to drug-abusing pedophiles, whom we now simply identify to be following an "alternative" lifestyle. Similarly, to brand a patient with "aberrant conduction" simply because electrical impulses travel a novel pathway from atrium to ventricle is an unfortunate stigma. Similarly, the connotations of the word "deviant" are to be avoided when describing the electrocardiographic axis of the heart ("left axis deviation").

The term "inferior myocardial infarction" has got to go. One can imagine the scenario: a patient returns home from the hospital and curious friends ask, "What kind of heart attack did you have?" To reply "My cardiologist says I had an *inferior* myocardial infarction" could lead to a cynical rejoinder. Better to use the anatomically and historically correct term, "diaphragmatic MI."

Valvular heart disease has not been spared from the application of demeaning and judgmental terminology. It is terribly insensitive to say that a cardiac valve is "incompetent" or "insufficient" simply because it leaks like a Washington health care committee. If the mitral valve leaks, thus allowing the retrograde flow of blood, would it not be better to say that the patient's heart is "retrograde mitral flow-enabled"? Similar terminology can be applied to any other leaky cardiac valve, as well as to the occasional "defects" found between cardiac chambers. If there is a hole between the left and right atria, the patient is "interatrial flow-enabled," a far better term than the current "atrial septal defect,"

The term "senile calcific aortic stenosis" is applied to older patients who have developed narrowed aortic valves impairing the outflow of blood from the heart. The term "senile" has connotations far beyond the cardiac system, and such patients would less stigmatized by applying the term "elder aortic flow-impaired."

Often, despite our best efforts, inotropic challenges and systolic impairments, combined with mul-tiple retrograde flow enablement syndromes, will lead to the decline in function of many of the body's organ systems. To apply the term "multi-system organ failure" smacks of the judgmental and paternalistic. Better to apply the term "metabolically challenged." This state all too often proceeds to the state of the patient's being "metabolically different," or "entropically enabled," in other words, dead. Patients should not proceed to this state of entropic enablement without the full benefit of modern, politically correct, humanizing, empowering, enabling, and sensitive cardiac terminology. "A heart is a heart is a heart. . . ."

The Guide to Politically Correct Cardiology

Instead of saying a patient has….	Instead say a patient is….
Heart failure	Inotropically challenged
Diastolic failure	Lusotropically challenged
Sick sinus syndrome	Chronotropically challenged; systolically impaired
Aberrant conduction	Alternative conduction
Left axis deviation	Left axis-enabled
Inferior myocardial infarction	Diaphragmatic MI
Hypercoagulable state	Rheologically impaired
Aortic (mitral) incompetence or insufficiency	Aortic (mitral) retrograde flow-enabled
Ventricular (atrial) septal defect	Interventricular (interatrial) flow-enabled
Poor surgical candidate	Cardiac medical therapy enabled
Senile aortic stenosis	Elder aortic flow impaired
Multisystem organ failure	Metabolically challenged
Dead	Metabolically different; entropically enabled

The Pop-Up Medical Thermometer

by Stephen Drew
AIR staff

This appeared in *AIR* 1:1 (January/February 1995).

Long-term nursing facilities have been wanting a cheap, simple way to tell when a patient is healthy enough to be discharged. Slow-recovering patients are often kept in bed for days or weeks longer than necessary. A new type of thermometer, which can be manufactured in bulk for less than four cents per unit, could save hundreds of millions of dollars annually in unnecessary medical expenses.

The pop-up medical thermometer is inserted into the skin or into existing body apertures. The inner cylinder pops up when the patient's fever subsides, indicating that it is time for the patient to go home. The technology was originally developed for the poultry industry. This is its first application to biomedicine.

When implanted transcutaneously in the belly, the pop-up thermometer can occasionally cause infections. The problem does not arise when the device is used rectally.

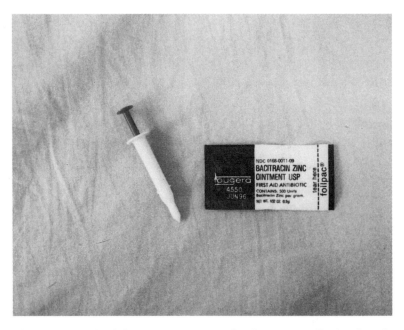

The pop-up medical thermometer uses technology originally developed for the poultry industry. The original version indicates when a chicken is fully cooked. Photo: Stephen Drew.

The Tomb of the Unknown Dentist

by A. J. Tuversen

This appeared in *AIR* 1:2 (March/April 1995).

In the hamlet of Lima, Ohio, near the west bank of the Root Canal, dentists from thirty nations have erected a memorial plaque at The Tomb of the Unknown Dentist. The significance of the number imprinted on the totem is, like the dentist, unknown.

An inscription at the base of the memorial reads:

> Rinse and spit,
> Rinse and spit,
> This won't hurt.
> My life is an amalgam
> Of one-sided conversations
> With open-mouthed people,
> Whose silent yearning need is great,
> Whose teeth are rotten.
> I hear their pain
> I feel their caries.[1]
> I patiently tell them all I know:
> I talk to them,
> I talk to them,
> I talk to them,
> I talk to them;
> Now they know the drill.

The memorial plaque at the Tomb of the Unknown Dentist. Photo: Stanley Rudin.

Note

1. Tooth decay.

The Dental Micro-Luger

Technology in the raw

by Walter Martin
SAS Institute
Cary, North Carolina

This appeared in *AIR* 1:2 (March/April 1995).

In dentistry, as in other fields, microtechnology is changing the way we work. This X-ray shows a case in point. The micro-Luger is being employed to break up an impacted wisdom tooth. In many cases, this makes the tooth's eventual extraction less traumatic to the patient.

The micro-Luger was developed in Germany, where it has still to be approved for clinical use. At present, licensing requirements to possess and use the device vary from nation to nation. In the US, the situation has taken on political importance in states where gum control is a hotly contested issue.

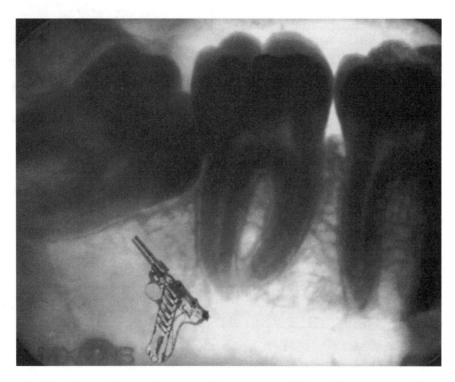

The dental micro-luger is becoming the treatment of choice for an impacted wisdom tooth.

Scientific Gossip

Contains 100% gossip from concentrate

compiled by Stephen Drew
AIR staff

These results are collected from various issues of *AIR*.

Dental Harmony

A survey of more than 16,000 dentists in the United States and Canada found that, no matter what the question, four out of five dentists agree.

Celebrity: The Inside Story

Mass market publishing companies are engaged in a bidding war over *X-Rays of the Rich and Famous*. The book is a compendium of images collected by former radiology technicians **Betty** and **Irving Francis**. Despite the threat of lawsuits from some of the patients featured in the book and by some of the hospitals at which the images were taken, the book is expected to take up residence on the best seller list. Publication is anticipated for the Christmas season.

Subatomic Homeopathy

Critics of homeopathy have always relied on a devastating argument: that homeopaths dilute their substances with so much water that typically not a single molecule of the original substance remains. Physicians and medical scientists commonly regard homeopaths as well-meaning crackpots. That may soon change, thanks to a theoretical breakthrough by **Deborah Linklater**, **Angus Dalziel**, and colleagues at The Center for High Energy Homeopathy, in Glasgow. Linklater and Dalziel say they will soon publish evidence that homeopathic substances are composed entirely of dissociated quarks and other subatomic particles, and that this neutralizes the water.

Paint for the Palate

Many children like to eat paint chips. Many children have nutritional deficiencies. These two facts form the basis of a new public health campaign. Los Angeles-based Nutripaint Corporation will soon bring to market an edible house paint that is non-toxic and vitamin-enriched. Their advertising campaign will be based on the slogan, "Let them eat paint."

Fat Discount

Restauranteurs are planning to take advantage of a new discovery in molecular genetics. **Fred Kern, Jr.** of the University of Colorado School of Medicine and **Richard B. Weinberg** of Wake Forest University's Bowman-Gray School of Medicine found that some people carry a gene that allows them to eat high cholesterol food without increasing their production of low-density lipoprotein (LDL), the carrier molecule that deposits cholesterol on artery walls. One major restaurant chain will soon urge consumers to take the simple blood test that shows whether they carry the gene. Those who do will be eligible for discount coupons for french fries and special high-fat hamburgers. (For pertinent information, see *Science News*, November 20, 1993.)

May We Recommend

Items that merit a trip to the library

compiled by Stephen Drew
AIR staff

These items were collected from various issues of *AIR*.

Tree-Induced Tropical Trauma

"Injuries Due to Falling Coconuts," by Peter Barss, *The Journal of Trauma*, vol. 24, no. 11, 1984, pp. 990–1. (*Thanks to James Barone for bringing this to our attention.*) The abstract reads in part:

> A 4-year review of trauma admissions to the Provincial Hospital, Alotau, Milne Bay Province, reveals that 2.5% of such admissions were due to being struck by falling coconuts. Since mature coconut palms may have a height of 24 up to 35 meters and an unhusked coconut may weigh 1 to 4 kg, blows to the head of a force exceeding 1 metric ton are possible.

Candied Birth

"A Prospective Study of Postpartum Candy Gift Net Weight: Correlation with Birth Weight," *Obstetrics and Gynecology*, vol. 82, 1993, pp. 156–8. (*Thanks to investigator Stephan Rössner for bringing this to our attention.*) The authors collected data from midwives who reported the weight of the chocolate boxes they obtained from grateful parents and the respective birthweight of the child. No correlation was found.

Bat Trauma in New York

"Impact of Yankee Stadium Bat Day on Blunt Trauma in Northern New York City," by S. L. Bernstein, W. P. Rennie, and K. Alagappan, *Annals of Emergency Medicine*, vol. 23, no. 3, 1994, pp. 555–9. (*Thanks to investigator Todd M. Mundle for bringing this to our attention.*) The authors sought to determine the incidence of blunt trauma in northern New York City before and after the distribution of 25,000 baseball bats at Yankee Stadium. They conclude that:

> The distribution of 25,000 wooden baseball bats to attendees at Yankee Stadium did not increase the incidence of bat-related trauma in the Bronx and northern Manhattan. There was a positive correlation between daily temperature and the incidence of bat injury. The informal but common impressions of emergency clinicians about the cause-and-effect relationship between Bat Day and bat trauma were unfounded.

Into the Digital Age

"The Digitally Obtained Stool Sample," (unsigned), *Emergency Medicine*, vol. 25, no. 16, December, 1993, p. 42. (*Thanks to Gauri Bhide for bringing this to our attention.*)

Psychiatric Reactions to Leeches

"Psychiatric reactions to leeches," W. A. James, R. L. Frierson, and S. B. Lippmann, *Psychosomatics*, vol. 34, 1993, 83–85. In one part of this study, the patient was instructed "to name the leeches as a means of empowerment." (*Thanks to Kerry Chamberlain for bringing this to our attention.*) The discussion makes recommendations for practice:

> During the initial phases of leech therapy, some patients may experience heightened anxiety. In [some] instances manifestations such as agitation, hostility and suspiciousness may herald the advent of serious emotional disturbances. . . . This could be important since the added stress of leech applications could exacerbate a depression with hopelessness and suicidal thinking.

Milk Abuse

"An assessment of the addiction potential of the opioid associated with milk," Larry D. Reid and Christopher L. Hubbell, *Journal of Dairy Science*, vol. 77, no. 3, pp. 672–675. *(Thanks to C. Robert Campbell for bringing this to our attention.)* The authors conclude that injections of milk are "not likely to become the focus of an addiction."

Think Before You Drink

"Milk and thought disorder," W. M. Bowerman, *Journal of Orthomolecular Psychiatry*, 1980, vol. 9, pp. 263. *(Thanks to Kevin Devine for bringing this to our attention.)*

Absorbing Attention

"Super effective diaper can cause confusion," A. Lavin, *Pediatrics*, vol. 78, no. 6, 1986, pp. 1173–4. *(Thanks to Gary Park for bringing this to our attention.)*

Flow From on High

"Urine and plasma proteins in men at 5400 m," D. Rennie, R. Freyser, G. Gray, and C. Houston, *Journal of Applied Physiology*, vol. 31, 1972, pp. 369–73.

Surgical Stylishness

"The Abdominal Zipper: A Surgical Surprise," V. Martinez-Ibanez, J. Lloret, and J. Boix-Ochoa, *Cirugia Pediatrica*, vol. 5, no. 3, July, 1992, pp. 182–3. It is of note that the concept of abdominal zippers was first proposed in writing by *AIR* co-founder Alexander Kohn in 1958, as a joke.

Rubbed the Wrong Way

"Masturbation Using Metal Washers for the Treatment of Impotence: Painful Consequences," A. Rana and N. Sharma, *British Journal of Urology*, vol. 73, no. 6, June 1994, pp. 722 ff. *(Thanks to investigators Vern Paxson and Ken Shirriff for bringing this to our attention.)*

Fetal Philosophy

University of Portland (Oregon) faculty members receive Arthur Butine Faculty Development Fund grants for their research projects. The list published in the January 30, 1995 issue of *Upbeat*, the university's weekly publication for faculty and staff, includes:

> Martin Monto, sociology, $4,500 to continue studying the meaning of childbirth.

Unanticipated Results: Significant Hips

"Adaptive significance of female physical attractiveness: role of waist-to-hip ratios," Devendra Singh, *Journal of Personality and Social Psychology*, vol. 65, no. 2, pp. 293–307. The author finds that

> Minor changes in WHRs of Miss America winners and Playboy playmates have occurred over the past 30–60 years and that college-age men find female figures with low WHR more attractive, healthier, and of greater reproductive value than figures with high WHR.

Mannequin Monthlies

"Could mannequins menstruate?" Minna Rintala and Pertti Mustajoki, *British Medical Journal*, Dec. 19–26, 1992, vol. 305, pp. 1575–6. *(Thanks to Doug Lindsey for bringing this to our attention.)* The authors explain:

> Mannequins that display clothes in fashion shops may influence women's perception of ideal weight. We investigated the changing shape of display figures over time and determined whether women of their size would have enough fat for menstruation. . . . A woman with the shape of a modern mannequin would probably not menstruate.

Unanticipated Results: Dog Bites Man

"Which dogs bite? A case-control study of risk factors," K. A. Gershman, J. J. Sacks, J. C. Wright, *Pediatrics*, vol. 93, no. 6, Jun 1994, pp. 913–7. *(Thanks to David Duffy for bringing this to our attention.)* The abstract reads in part:

> Conclusion: Pediatricians should advise parents that failure to neuter a dog and selection of male dogs and certain breeds such as German Shepherd and Chow Chow may increase the risk of their dog biting a nonhousehold member, who often may be a child.

Intercontinental Contraceptive Effect

"Geographic Tongue During a Year of Oral Contraceptive Cycles," J. Waltimo, *British Dental Journal*, vol. 37, no. 3–4, August 10–24, 1991, pp. 94–6.

Evacuation, Collapse

"The collapse of toilets in Glasgow," J. P. Wyatt, G. W. McNaughton, and W. M. Tullet, *Scottish Medical Journal*, vol. 38, 1993, p. 185. *(Thanks to Vidya Rajan for bringing this to our attention.)* The introduction reads in part:

> We describe three patients who presented during a period of six months with injuries sustained whilst sitting on toilets which unexpectedly collapsed.

Nemesis to a Comb

"The Uncombable Hair Syndrome with Pili Trianguli et canaliculi," A. Dupre and J. L. Bonafe, *Archives of Dermatological Results*, vol. 261, 1978, pp. 217–8. *(Thanks to James Rose and numerous other readers for bringing an entire body of related literature to our attention.)*

We welcome your suggestions for this column. Please enclose the full citation (no abbreviations!) and a photocopy of the paper.

AIR
Boys Will Be Boys

Research by and for adolescent males of all ages and sexes

by Katherine Lee
AIR staff

These items were collected from various issues of *AIR*.

All-Natural Power Generation

"Electricity Out of the Toilet Bowl," B. Miller, *Search*, vol. 25, no. 8, 1994, p. 246. *(Thanks to Paul Rattray for bringing this and the next item to our attention.)*

End Results

"A Survey of Hospital Toilet Facilities," A. F. Travers, E. Burns, *et al*, *British Medical Journal*, vol. 304, no. 6831, 1992, pp. 878–9.

Mandated Promiscuity

"Sexual promiscuity: knowledge of dangers in institutions of higher learning," R. D. Ebong, *Journal of the Royal Society of Health*, vol. 114, no. 3, June 1994, pp. 137–9. *(Thanks to Ken Shirriff for bringing this to our attention.)* The abstract reads in part:

> Students agreed that lack of financial support and of supervision from parents and teachers were among the causes of sexual promiscuity. The author concludes with a recommendation that a compulsory course on sexual promiscuity should be included in the syllabus in institutions of higher learning.

Ports of Entry

"Effect of ingested sperm on fecundity in the rat," R. A. Allardyce, *Journal of Experimental Medicine*, vol. 159, 1984, pp. 1548–53. *(Thanks to Barbara Piacente for bringing this to our attention.)*

Coffee Substitute

"Melatonin supplementation from early-morning auto-urine drinking," H. H. Mills, T. A. Faunce, *Medical Hypotheses*, vol. 36, no. 3, Nov. 1991, pp. 195–9. *(Thanks to Stephanie Faul for bringing this to our attention.)* The authors report that:

> Drinking one's morning urine ('amaroli') is a traditional practice of the yogic religion.... [This] ... restores plasma night-time melatonin levels due to deconjugation of its esters to melatonin. Exogenous melatonin ... may be the mechanism behind the alleged benefits of 'amaroli.'

Neuro-urinary Cartography

"Mapping of Brain Activity During Urination," talk delivered by Bertil Blok of the University of Groningen at the Society for Neuroscience annual meeting in November, 1995. *(Thanks to John Travis for bringing this to our attention.)* A handout described (among other things) a difficulty the researchers had to, er, face:

> An initial brain image was made while the men had a filled bladder prior to urination. A second recording was made when the bladder was empty. Nine of the fourteen volunteers were able to urinate under these artificial and difficult circumstances.

The Dark Recesses of Computer Science

"Designing efficient parallel algorithms on CRAP," Tzong-Wann Kao, Shi-Jinn Horng, Yue-Li Wang, and

Horng-Ren Tsai, *IEEE Transactions on Parallel and Distributed Systems*, vol. 6, no. 5, 1995, pp. 554–60. *(Thanks to William Korfhage for bringing this to our attention.)*

Bean Counters

"Further studies in flatometry," D. Fan, J. Tomlin, and C. L. A. Leakey, paper presented at the Bean Improvement Cooperative Meeting, which took place at the Marriott University Place, East Lansing, Michigan on October 25–28, 1995. *(Thanks to Bob Clark. for bringing this to our attention)*

Sticking Point

"Copulation as a possible mechanism to maintain monogamy in porcupines, *Hystrix indica*" Z. Sever and H. Mendelssohn, *Animal Behaviour*, vol. 36, no. 5, 1988, pp. 1541–2. *(Thanks to Wendy Cooper for bringing this to our attention.)*

Safe Seating

"The Gonococcus and the toilet seat," James H. Gilbaugh, Jr. and Peter C. Fuchs, *The New England Journal of Medicine*,, vol. 301, no. 2, 1979, pp. 91–3. *(Thanks to J. E. Charlton for bringing this to our attention.)* The article reads:

> . . . it is clear that toilet seats contaminated with purulent discharges containing gonococci may harbor viable organisms for several hours. But . . . [we find that] . . . contaminated toilet paper has greater potential as a direct source than do toilet seats.

AIR Vents

Exhalations from our readers

Note: These letters are collected from various issues of *AIR*. The opinions expressed here represent the opinions of the authors and do not necessarily represent the opinions of those who hold other opinions.

Stunning Images

The enclosed photographs of my collection of kidney stones are my only copies so please be careful because I have no others. I recommend that you use photograph #23 on your cover. It shows a kidney stone in the shape of Margaret Thatcher, the former Prime Minister of England, which is exceedingly rare. Alternatively, use photograph #9 which looks like a slipper.

B. T. Loess
Hamburg, Germany

Cooled Warmth

I was at first offended by your report on "The Pop-Up Medical Thermometer," a device adapted from the poultry industry and now used by hospitals to tell when a patient is ready to be discharged. But then I ran across this citation from the medical literature: "Colonic Removal of a 'Pop-Up Meat Thermometer' from the Sigmoid Colon," by R. G. Norfleet, G. Skerven, and H. T. Chatterton, *Journal of Clinical Gastroenterology*, vol. 6, no. 5, pp. 477–478. I apologize for my previous skepticism. Please keep up the good, er, work.

Lynn Pleister, Ph.D.
UCLA
Los Angeles, CA

Image Correction

Thank you for publishing the photographs of my collection of kidney stones. The photo on your cover shows what I had told you is a kidney stone in the shape of Margaret Thatcher, the former Prime Minister of England. I was mistaken. It was not a photo of a kidney stone. It was a photo of Margaret Thatcher.

B. T. Loess
Hamburg, Germany

Bibliophile and Meat Enthusiast

Jeffrey Moran's poem "Fifty Ways to Love Your Liver" should have included a reference to Philip Roth's book *Portnoy's Complaint*. Essentially, our hero uses a piece of raw liver as an erotic auto-stimulant. Well, he is my hero, anyway.

Eric Siegel
New York, NY

Cooking the Books

I have multiple chemical sensitivities. I am particularly sensitive to pesticides and printed (laser photocopied or printed, magazines, newspapers, some books) items. For many years I had to literally bake books before I could read them. I don't seem to have a problem with standard paperbacks, but larger softcover books, cookbooks for example, sometimes give me a rash. Do you have detailed information on how to bake a book? The heating process speeds up and promotes outgassing. I tried it with a romance novel today for about 1/2 hour at about 250 degrees, but I was only guessing. I'd rather have more explicit instructions before I set your magazine on fire.

H.D. Lamarck, M.D.
Mains, France

Licking AIDS

While driving to work, my wife and I heard a news report that researchers have discovered that saliva can counter the AIDS virus. Immediately, we concluded that this remarkable scientific discovery was probably the result of an over-worked, underpaid technician spitting into an AIDS culture in a moment of levity. But this discovery has led us to ask further questions: Given the frequency with which dogs lick their genital regions, does this make them less susceptible to the transmission of AIDS? If licking the genital area, before, after, or during intercourse reduces the transmission of AIDS, how do you disseminate this information without gaining the attention of the religious right? Realizing that researchers often avoid human testing due to the inability to obtain subjects, should they limit the number of volunteers for a test of oral sex as part of AIDS prevention?

Lawrence and Angela Collins
Auburn University
Marine Extension and Research Center
Mobile, Alabama

CHAPTER 8
Math and Models

Mathematics has an undeserved reputation for being dull and abstruse. Mathematicians have a generally well-earned reputation for being not so much absent-minded professors as absent-minded dressers. That is generally because their minds are on other, finer things such as truth, beauty, and money.

In the 1970s and 1980s a lot of government money was thrown into and plowed under an exciting, if ill-defined, concept called artificial intelligence. Mathematicians and computer scientists ran wild with money and ideas, at least for a while. "Achievements in Artificial Intelligence" presents a triumphant summary of their accomplishments.

Perhaps you have never given much thought to telephone numbers. Yihren Wu and Xiaohui Zhong have given too much thought to the subject. Their magnum opus is called "The Mathematics of Telephone Numbers."

Mathematicians also deal in truth, a commodity that is said to have little value in modern society. Joseph Cliburn, Andrew Russ, and their colleagues have mined the works of Bob Dylan for gems of mathematical value and truth. They present their findings in "The Value of Love, Using the Bob Dylan Model."

Simple math techniques can clear away a tremendous amount of clutter. Thanks to a book called *The Structure of Scientific Revolutions*, which was written by Thomas Kuhn and published in the early 1960s, the word "paradigm" has crept into the language of thousands of pompous reports and speeches. Bill Schweber's "The Paradigm Paradox" tracks the rise and fall of this pernicious word. One of our readers from built on Schweber's work to draw the obvious conclusion, as you will see in one of the letters in "*AIR* Vents." "*AIR* Vents" also contains a missive from one of our most prolific correspondents, Harold P. Dowd [with a "w," not a "u"].

Mathematicians either love words or love to compress them into symbols. "Mathematics—an Anagrammatical Tale" shows the poetical side of those who live and breathe numbers.

Mathematics is pure. It deals with ideal objects. Mathematicians, more than other scientists, speak of beauty and elegance. Some of their work is on display in scientist/supermodel Symmetra's ad for "SymmetriCal," and in the collection of photos from *AIR*'s annual swimsuit issue ("Scientific Swimsuit Sweeties"), and in Karen Hopkin's explanation of how she conceived and carried out "The Studmuffins of Science Calendar Project."

Advances in Artificial Intelligence

by Albrecht Grumme
Institut AI, Frankfurt, Germany

Fritz Schmelzeisen
Klinik AI, Hamburg, Germany

and Helmut Helmke
AI GmbH, Cologne, Germany

This appeared in *AIR* 1:3 (May/June 1995).

The Mathematics of Telephone Numbers

by Yihren Wu
Hofstra University
Hempstead, New York

Xiaohui Zhong
University of Detroit Mercy
Detroit, Michigan

This appeared in *AIR* 1:5 (September/October 1995).

In this paper we begin the study of a class of mathematical expressions of the form

$$x_1 x_2 x_3 - y_0 y_1 y_2 y_3 x_i, \; y_i \in \{0,1,2,\dots,9\}$$

Such expressions are generally known as **the phone numbers**. They involve the **subtraction** of a four digit number from a three digit number. These numbers are compiled annually and published by the local phone companies that were at one time associated with Bell Labs.[1]

Unlike the other research publications of Bell Labs, which are not widely accessible, the **phone book** is widely subscribed to. This may explain why there is no publication cost if one decides to publish his number in the phone book. However, if one decides not to publish, there is a non-publication cost. To encourage publication, there is no funding available to cover this cost, government or otherwise.

Our result is contained in a table that is too large (1000×10000) to be reproduced here. A portion is shown in Table 1. The complete table is available electronically by request.

Phone numbers are known to exist in a more complicated form, one of which defines the **long distance numbers**. Such numbers typically appear as

$$313 - 463 - 1645$$

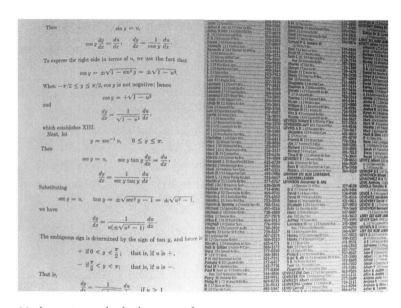

Mathematics and telephone numbers.

As it stands, the expression is ambiguous since subtraction is *not* associative:

$$(313 - 463) - 1645 \neq 313 - (463 - 1645)$$

An experiment performed on our Macintosh computer with a built-in calculator shows that the left-hand side equals –1895, while the right-hand side yields +1495. In fact, this experiment is not necessary, because the non-associativity of subtraction is a well-known fact.[2]

Table 1: The Telephone Numbers*

–	0000	0001	⋯	1645	⋯	9999
000	0000	–0001	⋯	–1645	⋯	–9999
001	0001	0000	⋯	–1644	⋯	–9998
⋮	⋮	⋮	⋱	⋮	⋱	⋮
463	0463	0462	⋯	–1182	⋯	–9563
⋮	⋮	⋮	⋱	⋮	⋱	⋮
999	0999	0998	⋯	–0646	⋯	–9000

* This is only a portion of the complete table.

An equivalent, although less common, way of expressing the long distance number is to utilize the more exotic operation of **multiplication**,

$$(313)\ 463 - 1645$$

In this multiplicative form, the product, (313) 463, is interpreted as the area, and 313, the first factor in the product, is the **area code**. It is, however, a misconception that a curve of long distance always fills an area; it only occurs in unusual circumstances, as reported by Peano.[3] One could argue that if the curve is closed, it becomes the boundary of an area, which is measured by the product in the long distance number.[4] This argument cannot be entirely correct, since a short distance curve can also bound an area, as illustrated by the following figure:

Area bounded by a long distance curve.

Area bounded by a short distance curve.

These complex geometric issues of phone numbers will be taken up in a forthcoming paper.[5] We are also in the process of tabulating a table for the long distance numbers. This table has dimensions 1000 × 1000 × 10000.

Acknowledgments

Part of this work was completed while one of the authors (Y.W.) was visiting the University of Detroit; we wish to thank X.Z. for her hospitality.

Notes

1. Cf. New York White Pages.
2. Private communication with Johnny, whose case has been reported in a study by M. Kline, *Why Johnny Can't Add: the failure of new math*, St. Martin's Press, New York, 1973. Although Johnny can't add, he and his generation are well educated in mathematical jargon such as this.
3. These are known as space filling curves, or the Peano curves, after G. Peano, "Sur une courbe qui replit toute une aire plane," *Mathematical Annals*, no. 36, 1890, pp. 157–60.
4. A different interpretation of this product is the cost. It has been observed that on occasion there is no cost if "0" is inserted at the beginning of the long distance number.
5. In preparation.

The Value of Love, Using the Bob Dylan Model

by Joseph Cliburn
Dept. of Institutional Research/Planning
Mississippi Gulf Coast Community College
Perkinston, Mississippi

Andrew Russ
Department of Physics
Penn State University
University Park, Pennsylvania

Tiny Montgomery
State Penn Center of Mathematics and Truck Driving
University Park, Pennsylvania

Zeke deCork
Shady Acres Old Folks Home and State University
Perkinston, Mississippi

This appeared in *AIR* 1:5 (September/October 1995).

Starting from a statement brought home by Bob Dylan [1965a], we estimate the value of Love using basic algebra of need [Mottram, 1965], perhaps some calculus, maybe a bit of the geometry of innocence [Dylan, 1965f], and a lot of wishful thinking.

The Limits of Love

We begin with the following assertion by Dylan [1965a]:

$$(Love - 0) / No Limit \quad (1)$$

using the expression on the record label in preference to the statement on the back cover [1965b], and taking a cue from the author's statement that it is a fraction [1965c].

Setting aside the question of whether the use of an expression here marks Dylan as an Expressionist, we set the expression equal to X, which is unspecified for the moment, and solve for Love:

$$X = (Love - 0) / No Limit \quad (2)$$

Thus:

$$(No Limit) X = Love - 0 = Love \quad (3)$$

where we've made use of the fact that for any A, A − 0 = A.

Thus Love = something times "No Limit." The traditional quantity that has no limit is infinite, thus we get Love is infinite, assuming that X is finite. If X is 0, we have 0 times infinity, which is indefinite.

Signs of Love

However, if X is negative, or "Less than Zero" [Costello, 1977], we get the result that Love is infinitely negative. This is perhaps enough negativity to succeed when gravity fails you [Dylan, 1965d], and will probably get the reader down. We may allow (no limit) to be negative, in which case we'll want either both X and (no limit) to be positive at the same time or both negative.

Other than the sign of X [Dylan, 1967a], however, there is nothing specified about it. If X is complex, then it has a real part that acts as above and an imaginary part, in which case (No Limit) times X is also complex, which makes Love both complex and partly imaginary [Whitfield-Strong, 196?]. Dylan himself has explored this idea extensively in later investigations [1975a, 1975b], with extensive revisions [1984, 1974/1993, various public presentations since 1975].

At any rate, we can conclude definitely [Anderson, 1982] that:

$$X = X \qquad (4)$$

We thus sum up by offering the following observations:

1. Love is infinite if X is finite.
2. Love is indefinite if X is zero.
3. Love is infinitely negative if X is negative.
4. Love is imaginary if X is imaginary.

Fractal Love is Problematic

There remain some questions regarding the appropriateness of using fractal mathematics to resolve these problems, e.g., "i accept chaos. i am not sure whether it accepts me" [Dylan, 1965e]. But we should also clarify that we are not putting infinity up on trial [Dylan, 1966] here. Love is, after all, just a four-letter word [Dylan, 1967b].

References

Anderson, L., 1982, "Let X = X," *Big Science*, (Warner Brothers, Burbank, CA).

Costello, E., 1977, "Less Than Zero," *My Aim Is True*, 2nd ed., (Columbia, New York, NY).

Dylan, B., 1965a, "(Love – 0)/No Limit," *Subterranean Homesick Blues*, (Columbia, New York, NY).

Dylan, B., 1965b, "Love – 0/No Limit," *Subterranean Homesick Blues*, back cover, (Columbia, New York, NY).

Dylan, B., 1965c, broadcast communication.

Dylan, B., 1965d, "Just Like Tom Thumb's Blues," *Highway 61 Revisited*, (Columbia, New York, NY).

Dylan, B., 1965e, liner notes, *Highway 61 Revisited*, (Columbia, New York, NY).

Dylan, B., 1965f, "Tombstone Blues," *Highway 61 Revisited*, (Columbia, New York, NY).

Dylan, B., 1966, "Visions of Johanna," *Blonde on Blonde*, (Columbia, New York, NY).

Dylan, B., 1967a, "Sign on the Cross," *Writings and Drawings*, (Random House, New York, NY).

Dylan, B., 1967b, "Love Is Just A Four-Letter Word," *Writings and Drawings*, (Random House, New York, NY).

Dylan, B., 1974/1993, "Tangled Up In Blue," *The Bootleg Series*, Vol. 2, (Columbia, New York, NY).

Dylan, B., 1975a, "Simple Twist of Fate," *Blood On the Tracks*, (Columbia, New York, NY).

Dylan, B., 1975b, "Tangled Up In Blue," *Blood On the Tracks*, (Columbia, New York, NY).

Dylan, B., 1978, " ," *Street Legal*, (Columbia, New York, NY).

Dylan, B., 1984, "Tangled Up In Blue," *Real Live*, (Columbia, New York, NY).

Mottram, E., 1965, *William Burroughs: The Algebra of Need*.

Whitfield-Strong, 196?, "Just My Imagination," as reviewed in R. Stones, 1978, *Some Girls*, (Atlantic, New York, NY).

The Paradigm Paradox

by Bill Schweber
Analog Devices, Norwood, Massachusetts

This appeared in *AIR* 1:1 (January/February 1995)

Scientific communicators are investigating the sudden drop-off in appearances of the noun "paradigm." Literature citations of the word (known as "citings of sightings") have decreased by 75% during the past year. The outlook for next year is even worse: the word has almost disappeared. More research is needed to chart the appearance/disappearance of paradigm since it first appeared in a widely-read book.[1]

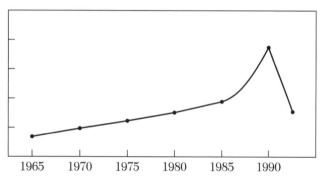

Appearances of the word "paradigm" in the general literature.

Hypotheses being proposed to explain this situation include global warming or other environmental causes, "wear out" and grinding down due to excessive use, or fashion, and user confusion as to what "paradigm" actually means.

A more advanced proposal from the Institute for Astronomical Linguistics is that the life cycle of a word parallels the life cycle of stars. When use of a word (or expression) grows slowly, it may become self-sustaining and eventually embed itself into the common language, resulting in a very long lifetime. In contrast, when usage flares up like a nova star, the mass of meanings that the expression must support becomes too great, and the expression suddenly collapses in on itself. In severe cases, it may become like a star transformed into a black hole, where it is never seen again.

Reference

1. *The Structure of Scientific Revolutions*, T. S. Kuhn, 2nd edition, University of Chicago Press, 1970.

Mathematics

An anagrammatical, if pointless, tale

by Alice Shirrell Kaswell
AIR staff

This appeared in *AIR* 2:4 (July/August 1996). It is one of a series of anagram tales that have appeared in *AIR*. Each line of this tale—except for the section headings—consists of an anagram for the word "mathematics."

Part 1: Tea Time

Me at MIT. Cash.

Me am sit chat.
Am chat: "It's me."
Ms: "It math ace."
"s'MIT math ace."
Same chat: MIT.
Is am math, etc.

Part 2: Play Time

Me: "MIT has cat."

Ms: "Ah, cat time!"

Time mash cat!
Me: "MIT cat. Ha!"
Me act. Sit cat.
Me: "Am sham cat."
"It am sham, etc."
Ms:"Him tame cats."
"Maim the cats."

Part 3: At Dinner

Me tact: "Is ham."

Ms am eat. Itch.
Me: "This am cat."
Me: "It's cat ham."
"Him's cat meat."

She: "Am MIT cat!"
Ham items act.
Hit Ms at acme.
Acts emit ham.
Ms emit cat. Ha!
Ms, at MIT, ache.

Part 4: Afterwards

That Ms am ice.
This mama, etc.

Versions of this advertisement began appearing in *AIR* in mid-1996.

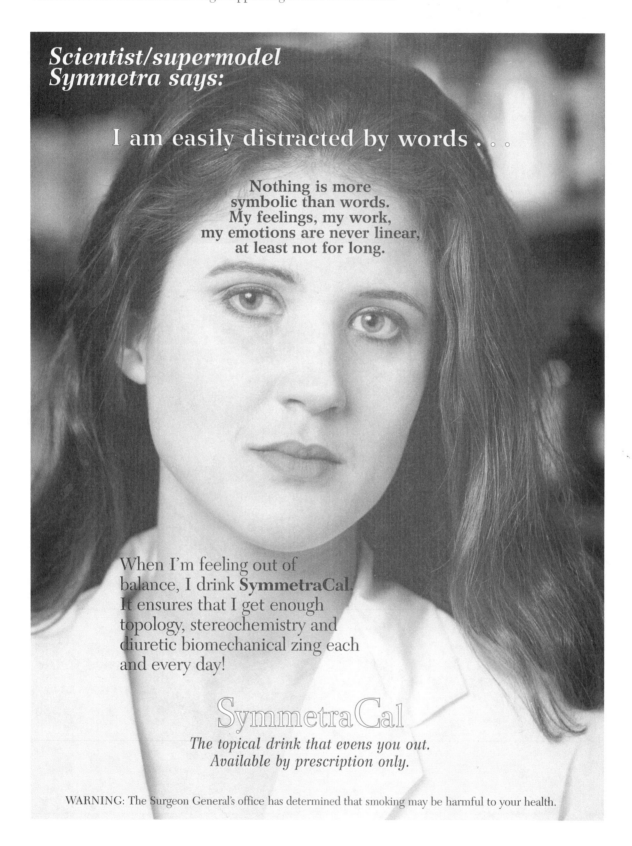

Annual Swimsuit Extravaganza

AIR's editors constantly roam the globe in search of nature's beauties. Here are two of their discoveries.

text by Lelivoldt Bruno
AIR staff

This appeared in *AIR* 2:2 (March/April 1996). Alert readers will notice that both of the models display their written handiwork elsewhere in this book.

Figure 1. Wet, wild, grammatical. The latest in a long line of southern science belles, Miriam Bloom has it all—a Ph.D. in genetics, the presidency of a major biomedical writing and editing firm (SciWrite^{sm}, based in Jackson, Mississippi), a new book (Understanding Sickle Cell Disease, from the University of Mississippi Press), a swimsuit and a snorkel. Here, Miriam emerges from the mighty Mississippi in midafternoon, making the transition from foamy frolic to fanciful footnoting. "I am engulfed in nature," she sighs.

Figure 2. Lean, bearded, confident. For more than 20 years, geologist Earle Spamer of the Academy of Natural Sciences of Philadelphia has been inclined toward intensive field work in the Grand Canyon. Here at the Little Colorado River, Earle demonstrates his method of total immersion in his work. Quicksand is no obstacle to his quest to empirically test the geological principle of the Angle of Repose, while disinterested colleagues stand idly by. "Barring rising flood waters, I expect to complete my definitive study within seven years," Earle boasts. "It is truly ground-breaking research."

The Studmuffins of Science Calendar Project

A *shameless personal testimony*

by Karen Hopkin, Ph.D.
Washington, DC

The idea popped into my head one afternoon while I was facing a 3 PM deadline and a blank computer screen. "Wouldn't it be a hoot if there were a calendar featuring real scientists posing like pinups?" Well-muscled and smiling in some sporty, sexy shots. It could be called...Studmuffins of Science!

Funny, huh? I certainly thought so. And so did pretty much everyone I told. In fact, the joke worked so well, it grew from a cute one-liner into a whole mini-routine. "Be a great way for me to meet guys," I'd continue. "I figure, with my ample talents, I'll be able to operate the spritz bottle at the photo shoots to keep the models moist."

Hahaha. Sure, it's all fun and games until *The San Francisco Examiner* calls, wanting to run a piece on your calendar, which, coincidentally, doesn't exist. "It's just a joke," I tried to explain. But it was too good a story to pass up.

So, the mass media made me do it. Now, two Studmuffins of Science calendars later, what have I learned? First, that most scientists, at heart, believe themselves to be studmuffins. I had *very* little trouble convincing my PhDs to pose. It's like they were just sitting in lab, waiting for the phone to ring. "A pinup calendar? Why, of course. I'll have my assistant bring in my Speedos at once."

Second, that all scientists really *are* nerds. But you can't always tell by just looking at them.

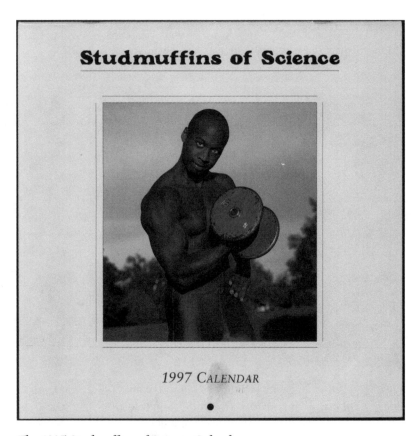

The 1997 Studmuffins of Science Calendar.

Finally, as a biochemist-turned-journalist-turned-calendar maker, I realize that I've made the biggest contribution I could have ever made to science by leaving the lab to pursue my dreams of, well, pursuing other scientists. It doesn't get much better than that.

Scientific Gossip

Contains 100% gossip from concentrate

compiled by Stephen Drew
AIR staff

This item appeared in 1992.

War Game Denial

The US Defense Department plans to dispute a soon-to-be-published report by Plaston University economist **Darlene Irons** that 42% of the funding for the Strategic Defense Initiative (popularly called the "Star Wars" missile defense system) was used to develop video games.

Ig Nobelliana
Words for the ages

"One of the other authors felt very strongly that we shouldn't acknowledge this."
—*A co-winner of one of the 1995 Ig Nobel Prizes.*

AIR Vents

Exhalations from our readers

Note: These letters are collected from various issues of *AIR*. The opinions expressed here represent the opinions of the authors and do not necessarily represent the opinions of those who hold other opinions.

Continental Digit Drift

Perhaps the authors of "The Mathematics of Telephone Numbers" would like to consider that their interpretation of "telephone number" needs to be more rigorously defined as "US telephone number." Their findings are not universal.

For instance, although Tokyo phone numbers used to be of that form, existing numbers acquired a "3" in front of the three-digit exchange number a few years ago. New numbers start with a "5," and it should be obvious to the reader that there is now room for further variations.

I have noticed that, whereas the numbers I have been assigned (3220–7475, 3220–5411, 3220–5412) all represent negative values, the numbers of my main clients (e.g., 3221–2031 and 5398–0784) represent positive values. A tentative model based on this particular set of data suggests that it might be related to cash flow.

Louise Bremner
Tokyo, Japan

Mathematical Surf

I'd like to propose an amendment to the Drake equation. [I understand there are two Drake equations; I'm talking about the first one, not the second.] The equation I'm talking about relates to the probability of finding intelligent extraterrestrials somewhere "out there" [out there being anywhere else but here]. The amendment of which I speak carries this calculation one step further: what are the odds of finding intelligent E.T.s that surf?

I propose adding to the Drake equation the factor, "1/x," where 1/x stands for the probability of an intelligent civilization evolving to the point where surfing is possible, if not commonplace. The revised Drake equation would thus consist of the original Drake equation multiplied by the "surfing correction factor," 1/x.

Serious comments on this subject are welcome. Jokers, however, are encouraged to keep their childish rejoinders to themselves.

Harold P. Dowd [with a "w," not a "u"]
Centerville, USA

P.S. Does anyone else have their own amendments to the Drake equation they'd like to suggest? If 12 people were to submit ideas, we could establish a 12-step program for updating said equation. This could be helpful for both extraterrestrials and E.T.s.

Shifty Explanation

Why did the usage of one word "paradigm" suddenly drop in 1990 after three decades of steady increase? Contrary to what Schweber, the author ("The Paradigm Paradox") contends, there is a simple explanation: in 1990 there was a paradigm shift.

R.L. Pramalal
Stanford University
Palo Alto, California

CHAPTER 9

Education, Scientific and Otherwise

Many of the people involved with *AIR* are or have been teachers. Most share the feeling that as long as you can rouse someone's curiosity, the rest of teaching is, if not easy, then at least possible. Without that curiosity, forget it. Now, there are some teachers who disagree. It is those teachers who have earned science teaching its sometimes sorry reputation.

Our general approach to teaching is to seduce people, be they kids or geezers or anyone in between, into looking at things that they believed to either booooooooring or impossible to understand. We believe in the by-hook-or-by-crook method. The letters and e-mail we get from teachers, parents and yes, students, hint that this method has a certain power and charm. (The letter presented in the "*AIR* Vents" section here makes the point eloquently.)

We print the "*AIR* Teachers' Guide" in every issue of the *Annals*, and occasionally reprint it in *mini-AIR* as well. If you are a teacher, you might find it useful. If you know a teacher, you might slip a copy to him or her.

How lively are today's students? Steven Rushen's "The Dead in the Classroom" will confirm your suspicions, hopes, and fears.

Some students need more seduction than others, pedagogically speaking. Dennis McClain-Furmanski's "A Mechanism for Getting and Keeping Students' Attention" will be of use when the need is great.

Good scientists come in all ages (as well as most shapes and sizes). Kate Eppers and Jesse Eppers, the authors of "Gummy Worm on a Sidewalk," were 12 and 10 years old, respectively, when they performed their research.

"*AIR*head Science Limericks" presents some of the many science limericks that our readers have sent in. We have an ongoing project for this. We have *many* ongoing projects, some others of which you shall glimpse in the chapter following this one.

What of the future of education? Anne Pamsun Hufnagle-Chang and Viktor Asa Gupta-Duffy's "Virtual Academia: Year 1 Report" will confirm or destroy every prediction you have ever read in the newspapers.

Teachers' Guide

This guide appears in every issue of *AIR*.

Three out of five teachers agree:

Curiosity is a dangerous thing,
especially in students.

If you are one of the other two teachers, *AIR* and mini-*AIR* can be powerful tools. Choose your favorite h*AIR*-raising article and give copies to your students. The approach is simple. The scientist thinks that he (or she, or whatever), of all people, has discovered something about how the universe behaves. So:

☛ Is this scientist right—and what does "right" mean, anyway?

☛ Can you think of even one different explanation that works as well or better?

☛ Did the test really, really, truly, unquestionably, completely test what the author thought he was testing?

☛ Is the scientist ruthlessly honest with himself about how well his idea explains everything, or could he be suffering from wishful thinking?

Kids are naturally good scientists.
Help them stay that way.

The Dead in the Classroom

by Steven Rushen
Penn State University
State College, Pennsylvania

This appeared in *AIR* 1:2 (March/April 1995).

The problem of when a person stops learning has received considerable attention. Many argue that people learn throughout their lives. Others assert that learning stops at an early age, and that any "learning" after that point is simply reapplying previous knowledge to fit a new situation. Many college professors believe that, for most people, learning stops sometime before a student's freshman year, giving further support to this second school of thought.

For my study I sided with the first school of thought. To an early morning freshman economics class of thirty live students, fifteen dead students were added and the effects were observed. After a full semester of careful study, the following observations were considered noteworthy. (See Table 1 for RIP[1] Coefficients.)

Attendance

On average, dead students are less likely to skip class than living students, especially on nice, warm days. Dead students had perfect attendance, were always in class early, and never left early (in fact they often stayed after and never complained when lectures ran long), unlike their living companions who had less than perfect attendance, were often tardy, and at times would leave early.

Behavior

On average, dead students were less disruptive than living students. Dead students are less likely to interrupt the instructor, be disrespectful, make noise, and ask irrelevant questions than their living counterparts.

Table 1: Measures of "Relative Individual Participation"

| Category | Mean Student RIP Coefficients[1] | |
	Living	Dead
Attendance	0.56	1.00
Behavior	0.40	1.00
Participation	0.12	0.13
Exam Scores	0.45	0.09

Class Participation

There was no discernible difference between living and dead students' performances in class discussions, responses to questions from the instructor, or when called to the chalkboard to solve a problem.

Exam Performance

This seemed to be the weakest point of the dead students. On average their scores were 30 to 40 points below the class mean. The effect this had on the grade curve was substantial, as it pushed the grades of all of the living students up to a B+ or better.

Conclusion

It is the author's opinion that dead students definitely have a place in the classroom. Their perfect attendance and exemplary behavior clearly illustrate their desire to learn. In three of the areas described they were at least the equal of, if not superior to, their living peers. While their performance on exams was poorer than that of living students, this cannot be taken as unwillingness to learn. The lower test scores could be due to low self-esteem, or to a misunderstanding, on the students' part, of general exam procedures. It is the author's opinion that in the near future "Outcome-Based Education" assessment may hold the key to overcoming this obstacle and give a better indication of the true learning ability of all students, vivacious or otherwise.

Note

1. RIP coefficients for Attendance and Exam scores are based on a straight percentage basis from performance in those respective categories. For Participation and Behavior this was based on both quantitative and qualitative measures of performance in these areas. Values of 1.00 equal a 100% or perfect performance, while 0.00 is 0%, or worst possible performance.

A Mechanism for Getting and Keeping Students' Attention

by Dennis McClain-Furmanski

College of Health Sciences
Old Dominion University
Norfolk, Virginia

This appeared in *AIR* 1:2 (March/April 1995).

Getting and keeping students' attention, particularly as the semester wears on, is always a problem. I offer to my colleagues the following method whereby I gain attention from the first moments of class, and regain it at any time during the year.

Purchase one of those "gourmet" candy canes from the stand most often found near the check-out counter of the grocery store. It must be plain white, no colors, no stripes. On the first day of class, break off an inch or so, peel off the cellophane, and head into the classroom with this candy hidden in your hand. Proceed to the chalkboard and pick up a piece of chalk. Write your name on the board with the chalk, and as you finish, switch the chalk with the candy. Turn around with the candy in place of the chalk, face the students, and while giving them an intense look of meaningful concentration, place the candy in your mouth and chew. The louder, the better.

During the next several minutes, it will occur to them that you have just eaten your chalk. As the recognition crosses their faces, you may at this time make note of the relative speed of cognition in this crop of students. Or, simply watch the facial contortions for your own amusement.

As the semester progresses, and attention wavers, you may from time to time stop in mid-sentence, walk to the board, grab a piece of chalk, and consider it closely for several seconds. You will have regained the full attention of the entire class.

Gummy Worm on a Sidewalk

by Kate Eppers, age 12
Marblehead Public Charter School
Marblehead, Massachusetts, and

Jesse Eppers, age 10
Horace Mann School
Salem, Massachusetts

This appeared in *AIR* 2:4 (July/August 1996).

We decided to do an experiment with gummy worms. We wanted to see how many people would step on a gummy worm, how many people would avoid the gummy worm, how many people would step on it without knowing it was there, and how many people would miss it completely.

Method

While visiting in North Conway, New Hampshire, we bought a bag of gummy worms from some place called "Fanny Farmer." We sat on a bench and threw a gummy worm into the middle of a sidewalk. We looked away casually, pretending we had no idea where the gummy worm came from.

Results

Adults and kids walked by, occasionally stepping on it. Some people would see it and give a puzzled look. A boy in a wheel chair rolled right over it without seeing it there. We tried not to laugh. Three teenage girls walked by it. The middle girl gazed down at it as she walked by. She let out a little gasp,

Figure 1: A gummy worm on a sidewalk.

and jumped over it. Then she laughed and said, "Hey, I thought that was real."

Conclusion

At the end of our experiment, we came to a conclusion that three out of five people will accidentally step on a gummy worm thrown on a sidewalk.

AIRhead Science Limericks

In 1995, we hesitantly announced (in *mini-AIR*) a new research project: The *AIR*head Science Limerick Compendium. The first limerick here came from reader Peter Olse, who claimed that he once used as it the entire answer to a final examination question: "Describe what you have learned in this course."

In Arctic and Tropical Climes,
The Integers, addition, and times,
Taken (mod p) will yield,
A full finite field,
As p ranges over the primes.

Over the next two years, the Science Limerick project elicited a sea of five-line fragments, some of them impressively bad. Here are some of the others. To the best of our knowledge, most of these were newly composed for the project, but with the authorship of limericks, one never knows for sure . . .

A violation of Sir Isaac was found
By Megan hurtling fast toward the ground.
She's not in smithereens
Because on trampolines
What goes down, must go up, then go down.

—Kevin Ahern

This is a limerick about a paper I submitted to "Physics Review E" entitled "Novel soliton solutions in Rowland ghost gaps":

In a periodic grating structure,
I claim Rowland ghosts should occur,
They have wriggles and bumps,
And travel over humps,
But the reviewer has yet to concur.

—Neil B.

My astronomy Ph.D. thesis in limerick form:

High-velocity clouds are found,
In disk galaxies to abound.
And although superbubbles,
Have given great troubles,
The fountain model is sound.

—Eric Schulman

A biology prof name of Caster
Had a project she knew would outlast her,
For it was most complex,
Aimed at changing the sex
Of *drosophila melanogaster*.

—Don Homuth

How Scientists Approach Limericks

The limericks inspired a brief note from editorial board member Jay Pasachoff as to what constitutes a properly formed limerick. That in turn spurred an intense and fascinatingly inconsequential debate on the proper rules of syntax, rhyme and content for Limericks. The entire set of commentaries, which one participant called "a Jesuitically Talmudic triumph of hyperbole and split hairs," will perhaps one day be shown to the public. Perhaps not. Here are two representative items. The second, as you will undoubtedly notice, is from the limerickaly driven Don Hormuth.

Please could you refrain from the sexist presentation "Da-da" for rhythmic measure? Dum (but not dumb) would be the appropriate choice of phoneme for non-politically-incorrect-speech (NPIS). NPIS is to be applied in conjunction with the rhythm method where possible. Deliberate misuse of the sex-determinant phrase "Da-da" leads to a harder life, and sleepless nights for scientists with young children.

—J. S. Notten

Written in haste, in immediate reaction to unjustified criticism.

There is always some guy with a rule
Who will claim that another's a fool
When his lim'rick won't scan

To a prearranged plan,
But forbear, it may *still* be a jewel.

—David Hormuth

But with that in mind, I hereby submit another, based on a TRUE event that occurred in 1965, when I was a lab assistant at North Dakota State University:

The zoology coed did squirm
At the lab quiz that ended the term.
When asked "What are tadpoles?",
(In the specimen bowls),
She wrote down "They are elephant sperm."

—David Hormuth

Our typographical mangling of his first name prompted Hormuth to write yet again:

I really hold no one to blame
For the fact of mispellling my name.
I submitted my rhyme
With the hope that, in time,
My name would turn flame into fame.

Big Boom in Limericks

Inspired, perhaps, by the Unabomber, reader Chris Marks composed three original scientific limericks with the common theme "Explosions of Various Sizes." They appear, for easy reference, in order of increasing magnitude of destruction:

A cautious young chemist named Mound
Was surprised (but not hurt) when he found
That A mixed with B
In the presence of C
Made a hole (ringed with dirt) in the ground.

Note: in this limerick, (R) represents the "registered" symbol

A scientist working at Sandia(R)
Found a way to make larger bombs handier.
The result of a test
In the desert Southwest
Turned the land close at hand even sandier.

Great minds have been known to recite,
Or in papers they publish, to write
That before time began

There occurred a Big Bang—
But the theory has never been quite completed.

Despite the kind offer of one reader from Singapore who offered to "brain" us if we published any more limericks, the project continued (and still continues). Here are a few more haphazardly selected limerickian efforts and commentary.

A research professor (Renee),
Cloned people from ape DNA.
The project went well,
Anyone can tell,
'Cause they're members of congress today.

—Frank Weisel

In Boulder, where often it snows,
NIST/JILA staff got high from lows.
A great celebration:
at last! condensation
according to Einstein and Bose!

—Walter Leight

Dr. Robert Stein sent us an essay about his adventures with pharmaceutical sales representatives. After lamenting that the drug companies no longer offer him free vacations to Hawaii, he concluded with this heart-rending flourish:

The rep from the drug company
Offers gifts that are no use to me.
Of that junk do me spare.
Gift me one year of *AIR*!
Till you do, go away! Let me be!

Mastodon, Mother, and Babe

Over a series of many months, our readers spun forth the stirring tale of the mastodon, the mother, and her babe. Reader Nancy White began the story, and *AIR* editorial board member Miriam Bloom composed the sequel. After that, others took up pen and pressed down keyboard to continue the tale. Here are several of their efforts cobbled together into what someone, somewhere, might call a story.

A Paleo hunter went walking,
Praying for prey worth the stalking.
She spied a huge mastodon

But hurriedly passed it on
'Cause the kid in her backpack was squawking!
—Nancy White

When she stopped to tend to the crying,
The mastodon upon her came spying.
So her sling she shot
(Gave it all she'd got)
But oops! 'Twas the babe that went flying.
—Miriam Bloom

The babe that went flying turned calm
Despite flying flung far from her mom.
But she wasn't in trance,
Was just loading her pants,
To besmirch the big beast with a bomb!
—Spencer Wright

To the mastodon's mouth, the babe flew,
and in its surprise it then threw-
up, as babes often will,
Its whole dinner, like swill,
and this made the poor mastodon spew.
—Peter Thorp

The poor airborne infant crash landed
'Tween the tusks of the mastodon, stranded.
The hunters soon tracked her
But failed to extract 'er
(This Pleisotcene group was short-handed).
—Heather M. Hewitt

The beast with the baby embedded
Must therefore be quickly beheaded.
But to climb o'er the trunk
Put the group in a funk;
Lo, those tusks and the task were both dreaded!
—Miriam Bloom

Readers are still sending us alternative histories of the babe, her mom and the mastodon. Despite insatiable public demand, we will spare you further details. At least for now.

Virtual Academia: Year 1 Report

by Anne Pamsun Hufnagle-Chang
Viktor Asa Gupta-Duffy
Department of Cognitive Administration
Milhouse College
Whittier, California

This appeared in 1993.

Virtual Academia is a virtual reality project designed to replace many costly aspects of today's universities.[1] This is a brief summary report on the first complete year of its operation.

The phrase "virtual reality" describes the use of computers to simulate objects and activities that occur in nature or in the imagination. In the Virtual Academia project, students, professors, lab equipment, classrooms, offices and dormitory facilities exist only as computer-based concepts.

Sixteen universities in seven nations replaced all or part of their traditional activities with the Virtual Academia system. A seventeenth university withdrew from the project because of equipment incompatibility.

Each university set its own system parameter values—admissions and hiring policies, grading curves, etc.—to conform with its own national and other administrative norms.

The Virtual Academia project's computer-generated classroom is in continual use. Here the virtual classroom acts as the simultaneous site of 223 virtual courses being taught to more than six hundred virtual students at sixteen universities. Photo: Alicia Ducovney-Lightpole.

Aggregate Results

Each class, seminar, research group and living unit in the sixteen universities was automatically balanced for gender, ethnicity and age to reflect the makeup of the larger society.

University general operating expenses were reduced by an average of 38%. Salary-related expenses were reduced by 54%, matching reductions in size of the faculty and support staff population. Student population was reduced by 83%.

Perhaps the greatest demonstrated benefit was that each university's numerical characteristics could be determined in advance, rather than having to be measured and explained after the fact. This consti-

tuted a significant reduction in administrative expense and activity.

Note

1. The project is funded by The Virtual University Network, a consortium of 91 universities and twelve nonprofit educational foundations. Sixteen universities were part of the test operations during year 1. An additional 42 universities will be brought on line during year 2. The remaining universities are scheduled to join the operation in year 3. For a complete membership list see TVUN Publication #146, *Organizational Membership of the Virtual University Network*, Hellgate Press, Berkeley, CA, 1996, $29.95.

Scientific Gossip

Contains 100% gossip from concentrate

compiled by Stephen Drew
AIR staff

This appeared in *AIR* 1:3 (May/June 1995).

Dumb Exaggeration

Word spread rapidly via the Internet last month: the National Intelligence Testing Service had announced incorrect results for many IQ tests administered between 1946 and 1985 reported incorrect results.

For many reported IQs between 101 and 130, the last two digits of the IQ number may have been were transposed. For example, someone who had a reported IQ of 120 may have an actual IQ of 102, and vice versa. We are trying to verify this.

AIR Vents

Exhalations from our readers

Note: These letters are collected from various issues of *AIR*. The opinions expressed here represent the opinions of the authors and do not necessarily represent the opinions of those who hold other opinions.

An Improbable Education

I want to congratulate you on the second issue and the very real impact it is already having. Yesterday (the day after I received it in the mail), my girlfriend's 14 year old son picked *AIR* 1:2 out of my pack (where my "current reading" is kept). He proceeded to read it intently cover to cover, especially "A Natural History of the Articulated Lorry."

This is a kid who, in the 6 years that I've known him, has betrayed relatively little interest in the various "educational" magazines (the usual: *Natural History*, *Smithsonian*, *National Geographic*, etc.) compared to the lure of Nintendo.

Not that there haven't been other factors behind his nascent interest in science. But we've had to fight the determined efforts of the public school system to get him to loathe science with a passion. The motivations may be different, but the actions are the same.

It's a long, uphill, and ongoing battle, that will doubtless continue until he's out of the clutches of the above-mentioned negative influences. *AIR* has given him a big boost. Thank you!!!

Mark Crispin
Bainbridge Island, Washington

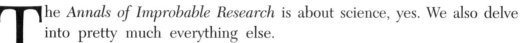

CHAPTER 10

Irrepressible Research

The *Annals of Improbable Research* is about science, yes. We also delve into pretty much everything else.

We have been told that, upon close examination, what we publish is as much about how people behave as it is about anything else. Of course it is. People are endlessly fascinating. People are strange. People are improbable. People are full of contradictions. Science is done by people who are trying hard, very hard, to be completely honest with themselves about what they know and what they don't. That is a doomed enterprise, but a noble one, more worthwhile perhaps than any other. What better theme could there be for any publication?

This final chapter pays an oh-so-fleeting visit to a few of *AIR*'s little adventures in this and that. Eric Schulman's "How to Write a Scientific Paper" gives an incisive between-the-lines explanation of the business of publishing scientific papers. For many scientists, this undertaking is one of the major things that affects the angle of ascent and the maximum height of their career paths.

"Furniture Airbags" is an exploration of technology transfer. It is also part of an ongoing safety crusade. Given time, we will make the world completely safe against any and all hazards. No one ever will suffer accident, illness, or—most important of all—inconvenience.

"Internet Barbie and the Time Caplet" tells the tale of how we went about celebrating the birth of *AIR*. Internet Barbie herself began as one reader's response to our call for that which would best symbolize the end of the twentieth century. Through the magic of trash recycling, we were able to turn that idea into reality.

Our Internet activities, especially our web site *HotAIR* and our monthly newsletter *mini-AIR*, are worth a book in themselves. For now, please be content with the snippets in the section called "Internet Adventures." And be sure to add yourself to the e-mail distribution list to receive *mini-AIR*.

Penultimately, there is "*AIR*head Project 2000." We were fascinated by the huge number of people and organizations who grabbed at the number 2000, all of them adding it to the name of some product, project, event, or something else that they wanted to market. Collectively, this clutching at 2000 is one of the world's most desperate, pathetic, and widespread uses of smoke and mirrors. We asked our readers to be on the watch for especially fatuous uses of 2000. For more than two years now these examples of idiocy have flowed to us by letter, by fax, and especially, by e-mail. The sheer volume of this mediocrity is its most impressive aspect. However, we would like to share with you here a few special gems.

The final, brief article, "With God in Mind" is unspeakably important.

How to Write a Scientific Paper

by E. Robert Schulman
Charlottesville, Virginia

This appeared in *AIR* 2:5 (September/October 1996).

Abstract

We (meaning I) present observations on the scientific publishing process which (meaning that) are important and timely in that unless I have more published papers soon, I will never get another job. These observations are consistent with the theory that it is difficult to do good science, write good scientific papers, and have enough publications to get future jobs.

Introduction

Scientific papers (e.g., Schulman 1988; Schulman & Fomalont 1992; Schulman, Bregman, & Roberts 1994; Schulman & Bregman 1995; Schulman 1996) are an important, though poorly understood, method of publication. They are important because without them scientists cannot get money from the government or from universities. They are poorly understood because they are not written very well (see, for example, Schulman 1995 and selected references therein). An excellent example of the latter phenomenon occurs in most introductions, which are supposed to introduce the reader to the subject so that the paper will be comprehensible even if the reader has not done any work in the field. The real purpose of introductions, of course, is to cite your own work (e.g., Schulman et al. 1993a), the work of your advisor (e.g., Bregman, Schulman, & Tomisaka 1995), the work of your spouse (e.g., Cox, Schulman, & Bregman 1993), the work of a friend from college (e.g., Taylor, Morris, & Schulman 1993), or even the work of someone you have never met, as long as your name happens to be on the paper (e.g., Richmond et al. 1994). Note that these citations should not be limited to refereed journal articles (e.g., Collura et al. 1994), but should also include conference proceedings (e.g., Schulman et al. 1993b), and other published or unpublished work (e.g., Schulman 1990). At the end of the introduction you must summarize the paper by reciting the section headings. In this paper, we discuss scientific research (section 2), scientific writing (section 3) and scientific publication (section 4), and draw some conclusions (section 5).

Scientific Research

The purpose of science is to get paid for doing fun stuff if you're not a good enough programmer to write computer games for a living (Schulman et al. 1991). Nominally, science involves discovering something new about the universe, but this is not really necessary. What *is* really necessary is a grant. In order to obtain a grant, your application must state that the research will discover something incredibly fundamental. The grant agency must also believe that you are the best person to do this particular research, so you should cite yourself both early (Schulman 1994) and often (Schulman et al. 1993c). Feel free to cite other papers as well (e.g., Blakeslee et al. 1993; Levine et al. 1993), so long as you are on the author list. Once you get the grant, your university, company, or government agency will immediately take 30 to 70% of it so that they can heat the building, pay for Internet connections, and purchase large yachts. Now it's time for the actual research. You will quickly find out that (a) your project is not as simple as you thought it would be and (b) you can't actually solve the problem. However—and this is very important—you must publish anyway (Schulman & Bregman 1994).

Scientific Writing

You have spent years on a project and have finally discovered that you cannot solve the problem you set out to solve. Nonetheless, you have a responsibility to present your research to the scientific community

(Schulman et al. 1993d). Be aware that negative results can be just as important as positive results, and also that if you don't publish enough you will never be able to stay in science. While writing a scientific paper, the most important thing to remember is that the word "which" should almost never be used. Be sure to spend at least 50% of your time (i.e., 12 hours a day) typesetting the paper so that all the tables look nice (Schulman & Bregman 1992).

Scientific Publishing

You have written the paper, and now it is time to submit it to a scientific journal. The journal editor will pick the referee most likely to be offended by your paper, because then at least the referee will read it and get a report back within the lifetime of the editor (Schulman, Cox, & Williams 1993). Referees who don't care one way or the other about a paper have a tendency to leave manuscripts under a growing pile of paper until the floor collapses, killing the 27 English graduate students who share the office below. Be aware that every scientific paper contains serious errors. If your errors are not caught before publication, you will eventually have to write an erratum to the paper explaining (a) how and why you messed up and (b) that even though your experimental results are now totally different, your conclusions needn't be changed. Errata can be good for your career. They are easy to write, and the convention is to reference them as if they were real papers, leading the casual reader (and perhaps the *Science Citation Index*) to think that you have published more papers than you really have (Schulman et al. 1994).

Conclusions

The conclusion section is very easy to write: all you have to do is to take your abstract and change the tense from present to past. It is considered good form to mention at least one relevant theory only in the abstract and conclusion. By doing this, you don't have to say why your experiment does (or does not) agree with the theory, you merely have to state that it does (or does not).

We (meaning I) presented observations on the scientific publishing process which (meaning that) are important and timely in that unless I have more published papers soon, I will never get another job. These observations are consistent with the theory that it is difficult to do good science, write good scientific papers, and have enough publications to get future jobs.

References

Blakeslee, J., Tonry, J., Williams, G. V., & Schulman, E. 1993 Aug 2, *Minor Planet Circular* 22357.

Bregman, J. N., Schulman, E., & Tomisaka, K. 1995, *Astrophysical Journal*, 439, 155.

Collura, A., Reale, F., Schulman, E., & Bregman, J. N. 1994, *Astrophysical Journal*, 420, L63.

Cox, C. V., Schulman, E., & Bregman, J. N. 1993, *NASA Conference Publication 3190*, 106.

Levine, D. A., Morris, M., Taylor, G. B., & Schulman, E. 1993, *Bulletin of the American Astronomical Society*, 25, 1467.

Richmond, M. W., Treffers, R. R., Filippenko, A. V., Paik, Y., Leibundgut, B., Schulman, E., & Cox, C. V. 1994, *Astronomical Journal*, 107, 1022.

Schulman, E. 1988, *Journal of the American Association of Variable Star Observers*, 17, 130.

Schulman, E. 1990, Senior thesis, UCLA.

Schulman, E. 1994, *Bulletin of the American Astronomical Society*, 26, 1411.

Schulman, E. 1995, Ph.D. thesis, University of Michigan.

Schulman, E. 1996, *Publications of the Astronomical Society of the Pacific*, 108, 460.

Schulman, E., Bregman, J. N., Collura, A., Reale, F., & Peres, G. 1993a, *Astrophysical Journal*, 418, L67.

Schulman, E., Bregman, J. N., Collura, A., Reale, F., & Peres, G. 1994, *Astrophysical Journal*, 426, L55.

Schulman, E. & Bregman, J. N. 1992, *Bulletin of the American Astronomical Society*, 24, 1202.

Schulman, E. & Bregman, J. N. 1994, in *The Soft X-Ray Cosmos*, ed. E. Schlegel & R. Petre (New York: American Institute of Physics), 345.

Schulman, E. & Bregman, J. N. 1995, *Astrophysical Journal*, 441, 568.

Schulman, E., Bregman, J. N., Brinks, E., & Roberts, M. S. 1993b, *Bulletin of the American Astronomical Society*, 25, 1324.

Schulman, E., Bregman, J. N., & Roberts, M. S. 1994, *Astrophysical Journal*, 423, 180.

Schulman, E., Bregman, J. N., Roberts, M. S., & Brinks, E. 1991, *Bulletin of the American Astronomical Society*, 23, 1401.

Schulman, E., Bregman, J. N., Roberts, M. S., & Brinks, E. 1993c, *NASA Conference Publication 3190*, 201.

Schulman, E., Bregman, J. N., Roberts, M. S., & Brinks, E. 1993d, *Astronomical Gesellschaft Abstract Series* 8, 141.

Schulman, E., Cox, C. V., & Williams, G. V. 1993 June 4, *Minor Planet Circular* 22185.

Schulman, E. & Fomalont, E. B. 1992, *Astronomical Journal*, 103, 1138.

Taylor, G. B., Morris, M., & Schulman, E. 1993, *Astronomical Journal*, 106, 1978.

Furniture Airbags

A glimpse at emerging technology

by Stephen Drew
AIR staff

This appeared in 1992.

It happens every day. Someone sitting in a chair leans too far backwards. The chair tips over. Severe head injury results.

The traditional method of protection is to wear a safety helmet (see Figure 1). However, this has always met with high consumer resistance. Thus was born the furniture airbag. The airbag technology was originally developed to protect automobile drivers in the event of a collision. Protective airbags are now being developed for chairs, couches and other domestic furnishings.

The chair/bag combination must guard against injuries that would happen if a person were to roll sideways after having tipped over backwards. This eventuality is not a major concern in the case of couches, of course, but it is a common problem with most chairs. One airbag per chair, it turns out, is not enough. At least two are needed to prevent the chair from rolling over. A three bag design is sufficient to preventing rolling, but may be economically undesirable—it raises the manufacturing cost far beyond current levels. Experiments with two large, softly inflated bags have been promising (see Figure 2).

Figure 1: A collision without airbags: The test subject, wearing a helmet to prevent head injury, collides with the floor after the chair has tipped over.

Figure 2: A collision with airbags: The rapidly inflated airbags protect the test subject, preventing head injury despite the fact that she is not wearing a helmet.

Internet Barbie and the Time Caplet

This appeared in *AIR* 1:2 (March/April 1995).

In late 1994, to celebrate the creation of *AIR*, we decided to bury a Tiny Time Caplet at MIT, and sponsored an essay contest to determine "What/who should be placed into the Time Caplet?" Suggestions included Ross Perot, Prince Charles, Newt Gingrich, Bill Clinton, Elvis, Carl Sagan, and top vote-getter Bill Gates. However, those individuals declined to be buried in the Time Caplet. The Time Caplet is unique among all Time Caplets buried in 1994 and all subsequent years, in that it contains NOTHING related to O. J. Simpson.

The contest winners received nothing, as this contest was devised by *AIR*heads. Here is the winning essay, by reader Donald Turnblade:

> I propose that a symbol embodying the properties of the Internet be enclosed in the proposed time capsule. It should represent the interconnectedness of the Internet, the human nature of the inhabitance of the Internet, the character of Internet communications, and the intellect of the Internet. Therefore, a half naked Barbie doll with fiber optic cables instead of doll hair would represent things fairly well. [Figure 1.]

Objects Buried in the Tiny Time Caplet

(All objects were carefully prepared with a trash compactor prior to insertion into the caplet.)

- Parking ticket
- Wad of gum chewed by a Nobel Laureate
- Bag of MacDonald's fries
- Madonna's pointy bra
- Internet Barbie
- Pentium Chip mounted on a Ouija Board
- "Penises of the Animal Kingdom" poster
- Vac-Man stretchable hydraulic doll (Figure 2)

Figure 1: The AIR staff constructed this working model of Internet Barbie.

- Cold War souvenir (packets of high-level radioactive waste)
- Bottle of Max Factor 2000 Calorie Mascara (representing *AIR*head Project 2000)
- The first issue of *AIR*
- Four Mighty Morphin Power Rangers as "Defenders of the Caplet"
- A bill for the cost of the Time Caplet
- A 1994 Ig Nobel Prize (a wax half brain mounted on a cheap wooden pedestal) (Figure 3)
- Three bacteria
- Star map (not of the celestial bodies, but of the homes of the stars in Hollywood)
- Can of Spam autographed by TV personality Robin Leach (Leach also sent special taped greetings and a recipe for chili)

Figure 2: Aided by scientist/supermodel Symmetra, Nobel Laureates Jerome Friedman (left), Richard Roberts (right), and Dudley Herschbach (right, partially obscured) prepare a Vac-Man doll for placement in the Time Caplet. A projection of nineteenth century architect Constant Desiré Despradelle oversees the procedure. Photo: Michael McCrory.

- Copy of Microsoft Windows
- Marilyn Vos Savant's soup bowl, spoon and napkin
- Oat bran and a running shoe
- Envelope of desiccant
- Potpourri

When the hole was dug, the *AIR* excavators discovered a time capsule that had been buried in 1914. It was unearthed and opened, The contents are listed here.

Objects found in the 1914 Time Capsule

- Parking ticket
- Hunk of (thoroughly) ripened cheese
- Book, "Einstein's Dreams," signed by someone named "Al"
- Cannister containing beard clippings
- Pack of Twinkies (still as tasty and fresh as the day they were made)
- Madame Curie's lipstick (phosphorescent)
- Schroedinger's cat
- One brown sock

Figure 3: As IQ recordholder and AIR editorial Board member Marilyn Vos Savant eats soup, fellow Ed Board member Dr. Thomas Michel explains his "Guide to Politically Correct Cardiology." AIR editor Marc Abrahams looks on. Above, marathon runner Bob Hersey presents his interpretation of AIR's logo, The Stinker. Photo: John Nanian.

- Scroll full of predictions:
    - PUBLISHING: Every businessman will have his own personal word-manipulator device—a coal-fired 65-horsepower typewriter.
    - POLITICS: In 1994—unlike today—our nation's leaders will be wise and educated men.
    - SPORTS: Babe Ruth will be the greatest slugger in Red Sox history.
    - ARCHITECTURE: a Chinese-American architect will build a glass pyramid in front of the Louvre Museum in Paris.
    - COMMUNICATIONS: There will be a World-Wide Web of communications. Every home will have its own telegraph, and every child will know the Morse Code. An automatic telegraph answering device will take messages when the family is not at home.
    - INTERNATIONAL POLITICS: An international League of United Nations will ensure that all nations live in peace and harmony. The headquarters for this international body will be in Sarajevo.

Ig Nobelliana

Words for the ages

"I'm not coming to your ceremony. I don't see how it will help me or my company."

> —*Jay Schiffman, upon being informed that he had won the 1993 Prize for Visionary Technology. Schiffman is the inventor of Autovision, a device that makes it possible to drive a car and watch television at the same time. He shared the Prize with the Michigan state legislature, which passed a law which made it legal to use Mr. Schiffman's invention.*

"This isn't like cold fusion—I can demonstrate it. Even with a pornographic videotape, you can drive in traffic, no problem."

> —*Jay Schiffman, in an interview published several months later in Omni.*

Internet Adventures

The jewel of the Improbable Research empire is the *Annals* itself. Every other month, a new issue of *AIR* is sprung upon the world. However, we always have material that is too timely— or just too tiny—to fit in the magazine. Some of this effluvia goes into *mini-AIR*, a free newsletter that exists only in electronic format (for info on how to add yourself to the *mini-AIR* distribution list, send email to <info@improbable.com>). It goes out over the Internet once a month, where many thousands of readers quietly, fervently redistribute it to friends, colleagues, and astonished superiors. In 1994, our webmaster and self-proclaimed Global Village Idiot, Amy Gorin, then of MIT, now of Stanford University, created the *AIR* web site (http://www.improbable.com), which we named Hot*AIR*. Here are a few items that have appeared in Hot*AIR* and various issues of *mini-AIR*. Many of them have been widely excerpted and reprinted in numerous publications and discussion groups on the Internet and elsewhere.

German Grammar

The official *AIR*head German language slogan, "Luft, luft, nichts als luft," which was announced last month, has become the source of disputatious consternation. A typographical error, compounded by the obscurity of the quotation source, infuriated or inspired many readers, especially Dr. W——f from Munich, who kindly mailed us an autographed sixteen volume set of his German grammar textbooks. Our thanks to him.

The Ig Nobel Peace Prize: Follow-up Investigation

Robert L. Park of the American Physical Society (APS) has done a follow-up investigation of the work which earned John Hagelin this year's Ig Nobel Peace Prize. Park's report appeared in his weekly APS newsletter, "WHAT'S NEW." It reads in part:

The [1994 Ig Nobel] Peace Prize went to physicist John Hagelin for his experiment to reduce crime in Washington, DC by the coherent meditation of 4,000 TM [Transcendental Meditation] experts. By coincidence, Hagelin was holding a press conference [on the day of the Ig Nobel Ceremony] to announce his final results. It was a data analysis clinic; violent crime, he proudly declared, decreased 18%! Relative to what? To the predictions of "time-series analysis" involving variables such as temperature and the economy. So although the weekly mur-der count hit the highest level ever recorded, it was less than predicted.

Valentine's Chocolate Survey: Phase I Results

1088 people participated in Phase I of the *AIR* St. Valentine's chocolate survey. We would like to thank each and every one of you—especially those who were kind enough to vote more than once. 73.27% of respondents were male; 25.42% were female; 1.03% could not or would not reveal their gender; 0.28% claimed to be both.

Aggregate preferences

62.34% prefer dark chocolate, 33.37% prefer milk chocolate, 1.03% voted for white chocolate, 1.78% were undecided. In addition, there was one vote each for light turkey, light beer, dark beer, marzipan and chocolate licorice. One person preferred the white chocolate in the form of a bunny. Two votes were of the X-rated variety. The answer "yes" was given twice.

We also received one Haiku.

There was one person who reported being born on Valentine's Day, and one person who reported being lactose intolerant and who considers milk chocolate to be the spawn of Satan. Seven people pointed out that we confused sex with gender and one gave the following reference: "Miss Manners'

Guide for the Turn of the Millenium," by Judith Martin, page 192." One person suggested an age factor be brought into the study, and two people recommended we examine the gender of the chocolate, i.e. with or without nuts.

As you may have noticed, we added white chocolate into the survey even though it was omitted from the original questionaire. This was suggested by a number of individuals, not all of whom voted for white chocolate. On the other hand, we recieved the following comment "White chocolate is an aberrant albino abomination which I'm glad to see you're ignoring."

The detailed results of Phase I:

Females

64.34% preferred dark chocolate
31.25% went for milk chocolate
01.10% for white chocolate
01.84% split their vote in favor of a dark milk chocolate
00.64% were undecided or other

Males

62.37% preferred dark chocolate
34.31% preferred milk chocolate
01.02% preferred white chocolate
10.2% split for dark milk
00.64% were undecided or other

Hermaphrodites

66.67% went for dark chocolate
33.33% chose milk chocolate

Unknown/Other

9.09% went for dark and milk respectively
27.27% opted for dark milk
54.44% were undecided or other

We leave it to educated readers to perform their own chi-square tests.

In Defense of Cindy Crawford

We must rise to defend the honor of a noted researcher, Cindy Crawford.

Recently we announced the premiere of the new *AIR* column "Cindy Crawford Discovers," which reports on the scientific efforts and achievements of supermodel Cindy Crawford (and her ilk) as documented in research journals such as *Vogue*, *Cosmopolitan*, *Elle*, etc.

The following day we received an e-mail inquiry from the magazine *Entertainment Weekly*, seeking an advance copy. We duly faxed a copy of the column, which concerns Cindy Crawford's recent achievements in chemistry.

A day later, *Entertainment Weekly*'s editors informed us that they consider Cindy Crawford's scientific achievements "too insubstantial" to report.

We must take exception. In our view, Crawford's lack of a Ph.D. in no way disqualifies her from conducting research with shampoos. To see this, one need look no further than her report concerning "a patented outrageous formula . . . enriched with pro-vitamin B5, silk protein and moisture-binding silicone." (The full text of Crawford's paper appears on page 11 of the January, 1995 issue of the research journal *Vogue*.)

Puzzling Predators

Thanks to our *AIR*head foolproof proofreading regimen, last month's issue of *mini-AIR* contained a grossly curious list of "African predators." Several thousand readers kindly wrote in to make sure we knew that tigers habitually dwell in Asia. Many also pointed out that (as put succinctly by investigator K. Hearn) "giraffes and wildebeest are generally not considered predators, unless one happens to be a leaf or a blade of grass." A reader identifying himself as "Art in Hollywood" was moved to the verge of poetry: "Ahhh those thrilling nature films showing the fearsome giraffe stalking it's wily prey—the leaf!"

Top Quark Tour (1995)

Congratulations and huzzahs to the physicists at Fermi National Accelerator Laboratory for finding evidence of that most elusive of subatomic particles, the Top Quark. We at *AIR* are arranging to acquire Fermilab's entire collection of top quarks and prepare them for public viewing. A travelling exhibition will stop in major cities around the world. The Top Quark tour schedule will be announced as soon as we solve the minor technical problems of how to preserve and mount the specimens.

Carning Error

The person identified in the February issue of *mini-AIR* as "John Carne" has requested that we identify him by his correct name, "James Carne." This James Carne resides in Amsterdam and is presumably male, though we make no firm claim in that regard. Nor do we make any specific statement as regards this individual's middle name or indeed on the question of whether this individual has a middle name, nor do we rule out the possibility that the J. (i.e., "James") Carne in question has more than one middle name. Please do not address correspondence on this matter to us, or to anyone named Carne, or to any resident of Amsterdam, or to anyone who can read. Thank you.

Top Quark Tour—Further Highlights

They are strange. They are charming. From Top to Bottom, our world-wide traveling exhibition of Top Quarks & Friends (from the Fermilab collection) has been drawing rave reviews. We are happy to announce that ticket prices have been reduced, thanks to the unexpectedly large run of spectators. Thrill as you infer that the Top Quarks decay before your eyes—in most cases before you even reach your seat. Every ticketholder receives a coupon redeemable for 40 million (!) free electrons, paradoxically delivered by the estimable Erwin the cat (Aye, there's the rub). A revised schedule of the remaining tour dates will be announced as soon as we receive the shipment of revised-formula formaldehyde plasma which will (the manufacturer promises us) enable us to preserve and mount the specimens.

Thin and Fat-Tailed Perpetuities –

We urge all *AIR*heads to attend the pair of lectures described here. (This information was announced in the Oxford University Gazette (April 27, 1995, no. 4361, vol.125). Thanks to Kate Morse for bringing it to our attention.)

> Professor C. Goldie of QMW London was scheduled to speak at Oxford University on June 13, 1995, at 3:30–4:00 P.M. on the topic "Thin and fat-tailed perpetuities." It was also planned that immediately after-

wards, Dr. S. Jacka was to speak about "Doob-like inequalities via optimal stopping."

This is part of a Midlands Probability Seminar. For more info, contact Dr. J. E. Kennedy.

Annals of Scientific Education

"Plagiarism is an Impossibility"

by Prof. Rebecca German
Biology Department
University of Cincinnati
Cincinnati, Ohio

[Note: This quasi-regular feature is a forum for those whimsical stories that detail the life of the mind, and that are experienced all too often by educators (formerly known as teachers).]

At a major university (is there any other kind?), students were asked to complete a writing assignment. This was to be an exercise in "critical thinking." Unfortunately, several students handed in essays that represented various degrees of intellectual dishonesty. These ranged from essays that were identical (except for the author's names) to essays with the same sentences rearranged. One pair of students, who turned in essays that were remarkably similar, filed a grievance against the professors who accused them of cheating. In a hearing before the grievance committee, it became clear that one of the students had written the essay and given it to the other. The student who admitted to copying her friend's essay did not believe that she had done anything wrong. She told the committee that, despite the identical sentences, in the two essays, she had only "looked" at the original. Finally, exasperated when the committee did not accept her story at face value, she stood up and said, "I only have to answer to two people: me and my God, and we both know I am right." One professor was heard to mutter under his breath, "what a shame only one of them is in the room to testify right now."

Manic-Depression Epidemic

As a contribution to public health, we ask you to help us increase awareness of the pandemic of manic depression. The disease is spreading rapidly. It was first brought to our attention recently, by a psychiatrist who is employed by a major American university. One of the university's football stars had recently been suspended for making ten yards and a cloud of dust out of his girlfriend. The football star had then been quickly reinstated to his rightful place on the team. The day after the reinstatement, a university psychiatrist explained on national television that the football player's actions had been caused by "the national epidemic of manic depression" that is "afflicting our elite athletes."

To our knowledge, this is the first public hint that manic depression might, just might, be a contagious disease. We urge you—especially if you are an elite athlete—to take suitable precautions against infection.

Re-Engineering for Everyone

This is a call for articles related to the topic of "Re-Engineering." The word "re-engineering" was coined several years ago at one of the world's outstanding technology institutes. Except for its meaning, the word is in no way related to the terms "reorganization," "downsizing," "organizational shrinking," "layoffs," or "eliminating large numbers of people." Re-engineering is known to be of vital importance: it has created financial and job security for many important consultants. For purposes of preparing research articles, and for all other purposes, "re-engineering" can be defined quasi-recursively:

> "Re-engineering is the process of instructing high-level administrators as to which people other than oneself are to be re-engineered."

More about Re-engineering

Several consultants have telephoned us (none of them used e-mail, because they wanted to make "personal contact") regarding the Re-Engineering project we announced last month. Each consultant offered his or her professional services, for a price, to help plan our special Re-Engineering issue. Several consultants also offered a second set of services should we desire their expert assistance not just in *planning* the issue, but in *designing* it. Three of the consultants offered a further level of service: arranging, conducting and evaluating a six-month series of focus groups to determine something or other.

We gathered the names and telephone numbers of all the consultants who phoned us, and then recommended them to each other as potential customers who are in need of re-engineering.

Global Village Deli Survey

Investigator Alius J. Meilus has identified a problem that must be addressed:

> I noticed that you have listed on your masthead a global village idiot. Since your village has a designated idiot, I was wondering if the village also has a good deli? After all, what is a village without a deli?

Spurred by investigator Mailus, we hereby announce a new survey, the Global Village Deli Survey. The survey consists of one question:

"Where can I find a good deli while visiting a particular neighborhood in the global village?"

Please send us any pertinent data that may be in your possession. If you send food, make sure it is fresh. The Global Village Deli Survey is yet another public service performed by *The Annals of Improbable Research*.

Ig Nobel Economics Update

One of the men who shared the 1995 Ig Nobel Economics Prize had more help than the public knew about. Robert Citron, the former treasurer of Orange County, California, consulted an astrologer and a psychic for investment advice. (This was reported in the Los Angeles *Times* last week.) The county lost approximately $1.7 billion. That's a lot of money, far more than a man of Citron's capacity could be expected to lose without skilled guidance. Which leads us to . . . [see next item]

Free Paranormal Abiltity Testing Service

As reported in the popular press, both the CIA and the KGB are hiring psychics. And they pay well.

If you are interested in whether or not you have paranormal powers, just sit down in a quiet corner and mentally send your name and address to Wojtek Bourbaki <bourbaki@neu.edu>, *AIR*'s resident ESP expert. If you don't receive notification from us within three days, you have no powers.

In the name of national security (any nation, any form of security) and economic growth (in either direction), *The Annals of Improbable Research* is offering this service free of charge.

Paranormal Spoon Incident

In the last issue of *mini-AIR*, we offered, free of charge, to test any reader who wished to know if he or she has paranormal powers. Testees were instructed to sit in a quiet corner and mentally send us their names and addresses. Alas, we had to terminate the testing program after readers in England and Israel reported a rash of bent spoons and then mentally lodged police complaints against us. We are now engaged in extra-cognitively presenting evidence to demonstrate that, whatever is bent or twisted, it is not the spoons.

PGP-Y

Our paranormal testing program has already had one commercial spin-off. Our engineers have developed a truly foolproof data security protocol. It is called PGP-Y—"Pretty Good Parasychology." The mechanism is simple. You imagine that you have transmitted data to someone; that person then imagines that he has received it. Using PGP-Y, any type of information can be transmitted over the Internet with complete security. The key is that the data is transmitted high over the net—so high that the data actually travels above the net rather than within it. The data is transmitted telepathically (and for those who distrust electronic funds, we also have a scheme for transmitting cash and gold plate telekinetically.)

Valentine's Bust

Just in time for your Valentine's Day dining pleasure, a gentleman named Dr. Blanton Cantellier of L'Institute d'Amour de Paris has developed a new method for slicing artichoke hearts into neat, tidy pieces without in any way peeling or cutting away the outer portions of the artichokes. Details were unavailable as *mini-AIR* went to press, but rest assured: we will continue to report on this heart breaking story.

The Strange Return of Karpook

Words, once written, have a way of reappearing. In May 1993, our editor published the following item in the magazine that he was then editing:

The "Teach Or I'll Leave" (TOIL) movement is gaining momentum. The movement was inspired by David Karpook, who as a Harvard undergraduate in the 1970's walked out of his Physics 12 class whenever the lecturer stopped making sense. In recent years the idea has spread to campuses across North America and thence to Europe, the Middle East and Africa. In recent months students at several Japanese technical universities have taken up the practice.

It is now 1996. Intrigued as always by the name "Karpook," we recently did a web search for that most euphonious of monickers. The search turned up, among other things, the entire passage that you just read. However, the words have somehow found their way into the middle of a book entitled "A Guide to the Philosophy of Objectivism."

Objectivism is a philosophy devised by the comic novelist Ayn Rand. The author of "A Guide to the Philosophy of Objectivism" is identified as "David King" from Milford, Wyoming. The entire book is online, and can be found at:

http://infosys.home.vix.com/pub/objectivism/ Writing/DavidKing/GuideToObjectivism/

The Karpook passage appears—verbatim—in Chapter 12, which is entitled "The Disastrous State of American Education." The Karpook story is presented as being fact—indeed King offers it as a crucial piece of evidence to support his argument.

This is a curious and lovely thing. When the Karpook item originally appeared (in 1993), it was clearly labeled "Scientific Gossip—Contains 100% Gossip From Concentrate." Yon bonny editor admits that he concocted it from a bare breath of fact: There was (and is) a David Karpook, and David Karpook did walk out of his Physics 12 class whenever he lost the thread of a lecture. But sad to say, there was no mass movement based on David Karpook's actions. Perhaps there should be.

Net Abuse: Announcing Project Whacko

During the past two weeks, we have received a surge of unsolicited e-junkmail. Much of it comes from banditos who use faked sending addresses that are difficult to trace. Inspired by Nobel Laureate Roald Hoffmann's theories of junk mail [*See page 17*], we announce the creation of Project Whacko.

Project Whacko is an ongoing research effort to induce electronic junkmailers to whack themselves out of existence. We invite you to send us simple schemes to help the e-vermin eliminate themselves. We will publish and disseminate the best of these techniques.

Here are the principles of Project Whacko:

1. The goal of Project Whacko is to prune the population of indiscriminate electronic junkmailers.

2. Project Whacko schemes will use judo/jujitsu principles redirect the e-junkmailers own evil actions back toward the putrid perpetrators.

3. Project Whacko schemes will themselves never involve the sending of indiscriminate e-junkmail.

Please send your responsible Project Whacko scheme to <bourbaki@neu.edu>.

Levels of Non-Meaning: The Recursive Hoax

What, indeed, is reality? Fed up with the persistence of pseudo-scientific pseudo-scholarly claptrap and gibberish, Alan Sokal submitted a load of intentionally utter nonsense to a "prestigious" "cultural studies" journal. 'Tis a wonderful piece of writing, indistinguishable from (and no less coherent than) the articles it mocks. The journal, *Social Text*, published this wonderfully moronic prose in its May '96 issue. Sokal, a New York University physicist, then wrote up the whole fiasco; he published his exposé in the magazine *Lingua Franca*. All this has been detailed in the general press.

But it may not be the whole story. We obtained a copy of *Social Text* and commissioned a panel of scholars (one of whom is a convicted felon) to read and deconstruct the text. The panel concluded— unanimously—that the other articles in *Social Text* are devoid of meaning and probably are themselves hoaxes. Thus Professor Sokal, thinking that he was cleverly showing up some rotten eggheads, was instead being suckered by a band of jokers more clever than himself.

So bravo, bravo, bravo to the deadpan merry old pranksters who call themselves "cultural studies scholars." Their many deadpan statements to the press in recent days are further triumphs in the grand dada style.

Note: Several months later, the editors of *Social Text* were awarded the 1996 Ig Nobel Prize for literature. Alan Sokal sent us congratulations.

Project *AIR*head 2000

compiled by Grigor Belfrey
AIR staff

These items were collected from various issues of *AIR* and *mini-AIR*.

With year 2000 fast[1] approaching, many[2] scientific, medical, engineering, legal, educational, governmental and marketing organizations are sponsoring research projects that that have "2000" as part of their names. Since 1994, our readers around the globe have been contributing to our list of studies, projects, and products. Every day, we receive anywhere from five to a hundred submissions. People tell us that they are appalled—and then goulishly fascinated—to see how very many people and organizations think they are being clever by using the magic number.

Four things inspired us to start this project:

1. the United Kingdom's **Education 2000** project
2. the US Department of Education's **Goals 2000** Initiative;
3. **2000 Flushes** toilet bowl cleaner; and
4. **Lever 2000 Soap**, which, according to its manufacturer, can be used to cleanse a human being's 2000 body parts.[3] This represents an important scientific discovery—namely, that the body has exactly 2000 parts.

*AIR*head Project 2000 (which we randomly also call Project *AIR*head 2000) is always seeking additions to the list. If sending actual products, please do not include any materials that have already been used and/or ingested.

Here is a tiny sample from the *AIR*head Project 2000 collection.

ITEM #9 (*submitted by investigator Steven Weller*)
Bassomatic 2000, *a fishing device.*

ITEM #9221-K7 (*submitted by investigator Kenneth A. McVearry*)

Salmon 2000, *a "fish initiative" by Onondaga County, New York.*

ITEM #5818 (*submitted by investigator Dudley A. Horque*)

Figure 1: 2000 Calorie Mascara, by Max Factor International, is item #3628 of the Project AIRhead 2000 collection. This specimen was submitted by investigator Deb Kreuze. Photo: Alice Shirrell Kaswell.

SCIENCE 2000, *event organized by the Scientific Suppliers Association of Australia in conjunction with the Victorian Wine Centre.*

ITEM #3280 *(submitted by investigator Daniel Rosenberg)*
Buns of Steel 2000, *an exercise video.*

ITEM #32-01 *(From the Dennis Geller Collection)*
Gluma 2000, *a dental material used with Pekafil, a universal dentin bonding resin.*

ITEM #6402-AB-4 *(submitted by investigator Alison, who apparently hath no last name)*
Europhalle 2000, *a tent at the semi-annual Urfahraner Market, offering beer and chicken to the accompaniment of Austrian folk music.*

ITEM #50388 *(submitted by investigator Robert Coontz)*
Teapot 2000, *"Specially commissioned by the Tea Council, Tea 2000 is a completely new way to enjoy the pure pleasure of tea. Beautiful to look at and superbly engineered, Teapot 2000 is uniquely designed to allow you to enjoy tea at the exact strength you like . . ."*

ITEM #1085-86
Domesday 2000, *a computerized network of databases detailing use, value, ownership, and boundaries of land in Britain.*

ITEM #LATX-0 *(submitted by investigator Amos Shapir)*
Condomat 2000, *a network of condom vending machines in Israel.*

ITEM #86-K *(submitted by investigator Jussi Karlgren)*
Kista 2000 (Coffin 2000), *a contemporary design coffin, white-lacquered fibreboard with light hardwood trimmings. Lining in cotton. Manufactured by Fredahls, in Aastorp, Sweden.*

Figure 2: BOB 2000, "the savings card that pays your bills," is item #0394 of the Project AIRhead 2000 collection. It is available from the First National Bank of Johannesburg, South Africa. The specimen was submitted by investigator Lynne Murphy of the Department of Linguistics, University of the Witwatersrand.

Figure 3: Arizona Odorless Garlic 2000 is item #21907 of the Project AIRhead 2000 Collection. This specimen was submitted by investigator Paul Jewell of Adelaide, Australia. Photo: Stephen Drew.

Notes

1. The rate of approach is approx. one year per year.
2. Approx. 2000.
3. The manufacturer originally stated this in its advertising, but has stopped doing so for reasons which we invite you to investigate

With God in Mind

by Alice Shirrell Kaswell
and Stephen Drew
AIR staff

Many people bristle at suggestions that religion can advance science or that science can be part of religion. We will now demonstrate that religion and science can profit from rubbing shoulders.

During the past decade, we have seen evidence that science and religion can touch in exciting ways. Physicist Stephen Hawking's book *A Brief History of Time* was a best seller, largely on the inspired strength of its final line:

> If we find an answer to that, it would be the ultimate triumph of human reason—for then we would know the mind of God.

Physicist Leon Lederman's book *The God Particle* also did well. Many in the publishing industry believe that the book's content, fine as it was, was irrelevant—that the power came from title. (We heard reports that Lederman wanted to call the book *The Goddam Particle* but bowed to his publisher's more enlightened wishes.)

The book you are reading at this moment is about science. It is an interesting book, perhaps an important book. With our current limited understanding of the universe, we as human beings cannot know who will read this book. Perhaps God will purchase a copy, or will receive a copy sent by a worshipful admirer. The important question is: will God enjoy this book? If we find an answer to that, it would be the ultimate triumph of human reason—for then we would know the mind of God.

AIR Info

mini-*AIR* is a free monthly electronic supplement to *AIR*.
To subscribe, send e-mail to <LISTPROC@AIR.HARVARD.EDU>
The body of your message should contain only the words:
SUBSCRIBE MINI-AIR MADAME CURIE
(You may substitute your own name for that of Madame Curie.)

Up-to-date news and schedules: <info@improb.com>
  or   <info@ improbable.com>
AIR **and Ig Nobel on the web:** www.improb.com
  or   www.improbable.com
AIR **bits on USENET:** clari.tw.columns.imprb_research

Get in Touch With Us

We do read all the mail, paper and electronic, that you send. However, we receive far more correspondence than we can answer. Please enclose a SASE if you need a reply to your printed mail. Please include your e-mail address (if any) in all printed correspondence. Here is our address:

 Annals of Improbable Research (AIR)
 P.O. Box 380853, Cambridge, MA 02238 USA
 (617) 491-4437   FAX: (617)661-0927
 http://www.improb.com
   or   http://www.improbable.com
 Editorial: <marca@wilson.harvard.edu>
 Subscriptions: <air@improb.com>
   or   <air@improbable.com>

Guidelines for Authors

We are unable to acknowledge receipt of printed manuscripts unless they are accompanied by a self-addressed, adequately stamped envelope.

AIR publishes original articles, data, effluvia and news of improbable research. The material is intended to be humorous and/or educational, and sometimes is. We look forward to receiving your manuscripts, photographs, X-rays, drawings, etc. Please do not send biological samples. Photos should be black & white if possible. Reports of research results, modest or otherwise, are preferred to speculative proposals.

Keep it short, please. Articles are typically 500–2000 words in length. Articles intended for mini-*AIR* should be much, much shorter. Please send two neatly printed copies. Alternatively, you may submit via e-mail, in ASCII format.

Annals of Improbable Research

ISSN 1079-5146

The journal of record for inflated research and personalities

You've read and heard about *AIR* in *Nature, Science, New Scientist, Scientific American,* the *New York Times,* the *Wall Street Journal,* the *Washington Post,* the *Times of London, Die Zeit, Haaretz,* on NPR, the BBC, ABC News, NBC News, CNN, C-SPAN, and all over the Internet. *Wired* magazine said that "*AIR* is one of the finest contributions to western civilization."

Now—for your own peace of mind—subscribe!

Send payment to:
AIR
P.O. Box 380853
Cambridge, MA
02238 USA

617-491-4437
FAX: 617-661-0927
<air@improb.com> or
<air@improbable.com>

RATES (in US dollars)		
	1 year	2 years
USA	$23	$39
Canada/Mexico	$27	$45
Overseas	$40	$70

✄ -

My name, address, and all that:

Name: _____

Addr: _____

Addr: _____

City:_____ State_____ ZIP: _____

Country: _____

Phone: _____

FAX: _____

E-mail: _____

Total payment enclosed: _____
Payment method:

_____ Check (drawn on US bank) or int'l money order

_____ MasterCard _____ Visa _____ Discover

CARD #: _____ EXP. DATE _____

Index

(* indicates that index listing has a star beside it)

NOTE: Because of software problems, some of the following index listings may be accurate.

Abduction by space aliens, probability of, 85
Academia, virtual, 182
AIR Vents (letters), 89, 104, 134, 157, 172, 184
AIR, address of, 3, 202
AIR, improbable history of, 1
Airbags, furniture, 189
Altman, Sidney, 29
Apple, Fiona, spectrographic comparison of oranges
    with, 358
Apples, spectrographic comparison of oranges with, 93
Arizona Health Sciences Center, 39
Artichoke hearts, 197
Artificial Intelligence, 161
Ask Symmetra, 97
Asses, 132
August, George, and collaborators, 79

Baerheim, Anders, 35
Bakkevig, Martha Kild, 37
Baltimore, David, 24
Barbie, Internet, 190
Barings Bank, 37
Barnard, Neal, 45
Barney, taxonomy of, 107
Baseball, 66, 152
Bassa, Ivette, 43, 71
Batman * , 24
Batra, Ravi, 41
Beano, 44
Beaumont, Robert, 38
Beer and chips, merits of, 29
Benveniste, Jacques, 44
Beverly Hills, 38
Bijan Fragrances, Inc., 37, 38, 121
Billiards, 21
Biosphere, 131
Blackford Hall, dining review of, 128
Bloom, Miriam, 169
Bolts, 308
Boo-boos, medical effects of kissing, 142
Bos, Nathan, 134
Bourbaki, 196
Bowel movements, military, 39
Bower, Doug, 43
Boyle, Michael, 37
Boys Will Be Boys *, 155
Breakfast cereal, soggy, 37, 38
Bremner, Louise, 172

Brown and Williamson (tobacco company), 35
Buebel, Marcia, 37
Busch, David, 37
Butterfly effect, evidence of, 60

C-SPAN, 31
Campbell, James and Gaines, 41
Campbell, William *, 35
Carberry, Josiah, 44
Cardiology, politically correct and/or approved, 146
Carne, John or perhaps James or possibly J.
Cellulose fibers, chucked, 114
Cereal, soggy, 38
Chalk, argument in favor of eating, 177
Chanteuses, whining, 399
Chaos, evidence of butterfly effect and *, 60
Charles, Prince, 427
Cheese, in Switzerland, 68
Chemistry & Industry (science journal), 34
Chile, national pride of, 39
Chips, potato, aerodynamics of, 75
Chirac, French President Jacques, 35
Chocolate survey, 193
Chorley, David, 43
Chung, Mary, respect for among scientists and food
    critics, 400
CIA, 196
Citron, Robert, 37, 196
Clemens, Samuel and Roger, curveballs compared and
    contrasted, 361
Clinton, Bill, 101
Cold Spring Harbor, review of dining hall of, 128
Collins, Lawrence and Angela, 158
Colonoscopy, world records, 427
Comic books, 24
Corning, Inc., 40
Cow suit, woman in, 33
Cox, Courteney, 103
Crab, photo of sad South African specimen of, 113
Crawford, Cindy, 102, 194
Crick, Francis, 122
Crop circles, 43
Curie, Marie, theatrical career of, 379

Da-da, 179
Danger, Nick, 666
Dart, Richard, 39, 40, 49
Darwin, Charles, alternative to, 51

David Jacobs, 42
Davila, Jan Pablo, 39
Dedication, vii
DeFanti, Paul, 44
Deli, global village, 196
Dental floss, 38
Dental Micro-Luger, 150
Derek, Bo, 38
Diana, princessness of, 427
Dickinson, Emily, pharmacokinetic research of, 404
Diet, recipes for disrupting, 421, 445, 492, 506, 584
Dietrich, Marlene, patents registered to, 373
Dingell, US Representative John, 419
DNA cologne, 37, 38, 121
Doornail, death criteria for, 137
Dover, effects of chlorine bleach on white cliffs of, 462
Dowd, Harold P. (with a "w," not a "u"), 172
Dull lecture, how to tolerate, 16
Dun Dun noodles, 400
Dylan, Bob, mathematical legacy of, 164

Eclaireurs de France, 43
Einstein, Albert, average number of hours of sleep per
     night, 432
Electric shock treatment for rattlesnake envenomation,
     failure of, 49
Entertainment Weekly, 194

Fatheads, 21
Featherstone, Don, 36
Fermi National Accellerator Lab, 194
Feynman, Richard, and adventure of the flying hookworm,
     361
Finegold, Leonard X., 85, 134
Flamingos, pink, 36
Foot odor, 43
Football, American, 196
Four, chapter, 59
Franklin, Aretha, 444
Fresh fruit, inability of submarine chef to prepare well,
     128
Friedman, Jerome, 45, 190
Fukuda, M., 43

Gardner, Martin, 4
Gates, Daryl, 43
Genco, Robert, 35
Georget, D.M.R., 37
German grammar, 193
German, Rebecca, 195
Glasgow, Texas State Senator Bob, 39
Glashow, Sheldon, 33, 45, 71
Glassware, kinetics of inactivation of, 8
Glioblastoma, that resembles Little Bird, 140
Goble, George, 35
God, 201
Golden Fleece Awards, 33
Goldie, C., 195
Gonorrhoea, transmission of through an inflatable
     doll *, 46
Goodbye, goodbye speech, 32
Gorbachev, Mikhail, as antichrist, 42

Gorin, Amy, 193
Graham, Robert Klark, 44
Grail, holy, 63
Grand Canyon, 169
Grass, cyclic growth of, 120
Grayson, Donald, 119
Greenland, sexually transmitted disease in, 46
Gustavson, Richard, 39, 49

Hagelin, John, 39
Harvard Medical School, 42
Hats, ugly, 486
Hawking, Stephen, 201
Herrnstein, Richard J., 100
Herschbach, Dudley, 19, 33, 45, 99, 191
Hersey, Bob, 191
Hieroglyphs, 125
Hoffmann, Roald, 17
Holmes, Sherlock, 27
Homeopathy, 44
Hubbard, L. Ron, 39, 41
Hurley, Elizabeth, 102
Hwang, Stanley, 388

Ig Nobel Prize Winners, 35
Ig Nobel Prizes, listing and history of, 31
Incident, embarrassing, 406
India, concise biographies of every citizen of, 509
Inflatable doll, transmission of gonorrhoea through *, 46
Insect, largest observed, 117
Institute of Food Research, 37
Institute of Science, Technology and Public Policy, 39
Intelligence quotient (IQ), 99, 183
Internet adventures, 193
Irreproducible to Improbable, history of, 6

Jacobs, David, 42, 85
Jacobson, Cecil, 43
Japanese evolution, 51
Japanese Meteorological Agency, 39
Jell-O, 43
Johnson, James, 35
Junk mail, electronic, 197
Junk mail, snail, 17

Kaiser Wilhelm, 111
Kanda, F., 43
Karpook, David, 197
Keio University, 37
Kervran, Louis, 41
KGB, 196
Klein, Calvin, 102
Kleist, Ellen, 36, 46
Kligerman, Alan, 44
Knowlton, Jim, and penises, 42
Kohn, Alex, 4, 6, 8
Korsnes, Norwegian Honorary Consul Terje, 40
Kyle, Thomas, 43

Lamarck, H. D., 157
Larrsen, Anders, 104
Larsen, Jamie, 104

Leacock, Stephen, 401
Lederman, Leon, 201
Leeches, 35, 152
Leeson, Nick, 37
Lehrer, Tom, 306
Lightman, Alan, 41
Limericks, 179
Lint, 132
Lipkin, Harry, 4, 6, 89
Lipscomb, William, 27, 33, 45
Liver, 50 ways to love, 141
Lloyds of London, 43
Loess, B. T., 157
Lopez, Robert, 41, 47
Lorillard (tobacco company), 35
Lorry, articulated, natural history of, 117, 184
Love, value of using Bob Dylan model, 164
Luak coffee, 37

Mack, John, 42, 43, 85
Maharishi University, 39
Malone, Lois, 32
Manic-depression, 195
Marx, Julius H., 444
Mastodon, 180
Mathematics, anagrammatical transform, 167
Mathematics, pointless, 167
Matthews, Robert, 35, 36
May We Recommend °, 88, 132, 152
May, Sir Robert, 34
McKinnon, Catherine, views on asexual reproduction, 348
Meyrieres Cave, 43
Michel, James, 146, 191
Michigan, state legislature of, 42
Mickeymouse Gene, 116
Microscopy, Xerox enlargement technique, 95
Milken, Michael, 44
Mini-creatures, 51
Mitchell, Tim, 40
Mites, ear, 47
Moi, Harald, 36, 37, 46
Monacles, Victorian women's sexual obsession with, 412
Mondocentrism, 81
Monet, discrimination by pigeons of paintings by, 37
Moon, fetal man in the, 145
Mozart, W. A., 99
Murray, Charles, 100
Murray, Joseph, 39

Nakajima, K. Ohta, T., 43
Nakata, O., 43
Nanotechnology, 72
National Endowment for the Arts, 43
National Public Radio, 31
Nematodes, 125
New Jersey, 503
New Yorker, 4
Newell, Kenneth, 41
Nicotine, non-addictiveness of, 35
Nielson, Ruth, 37
Nixon, Richard, as organic chemist, 496

Nobel laureates, interviews with, 15
Nolan, James, 41, 42, 144
Norwich, England, 37
Nuts, 308

Okamura Fossil Laboratory, 51
Okamura, Chonosuke, 36, 51
Okamura, Chonusuke, relationship to Okamura Fossil Laboratory, 51
One, page, 1
Orange County, California, 37
Oranges, spectrographic comparison of apples with, 93
Oranges, spectrographic comparison of Fiona Apple with, 358
Oregon State Health Division, 41
Oslo, Norway, 37
Oxygen, liquid, use in cooking, 35

Pakzad, Bijan, 37, 38
Paltrow, engagement to Pitt, 437
Paradigm paradox, 166
Paradox, paradigm, 166
Paris, rainfall in, 60
Park, Robert, 193
Parker, R., 37
Particle accelerator, Switzerland, 66
Paul Williams, 41
Pauling, Linus, 25, 94, 122
Peanut butter, 79
Pencils, merits of versus pens, 19
Pendulum, Pittsburgh and the, 435
Penis, zipper-entrapped, 41,42
Penises of the Animal Kingdom, 43
Pens, merits of versus pencils, 19
Pepsi-Cola, 42
Periodic table, politically correct, 101
Perseverence, 299
PGP-Y, 197
Pheromone coupon, 138
Philip Morris (tobacco company), 35
Phonies, 21
Physicians' Committee for Responsible Medicine, 45
Pitt, abbreviation for Pittsburgh, 436
Pitt, engagement to Paltrow, 437
Pleister, Lynn, 157
Poets, obscurity of certain British, 304, 334, 338, 372, 395, 419, 444, 462, 489
Pop singers, high-voiced male, irritation caused by, 399
Pop-up medical thermometer, 148
Popeil, Ron, 42
Potato, alternate method for taking a, 428
Pramalal, R.L., 172
Proxmire, US Senator William, 33
Purdue University, 35

Raclette, laser cheese, 68
Rapeseed, 130
Rattlesnake envenomation, failure of electric shock treatment for °, 49
Re-engineering, 196
Rectal objects, 37

Refrigerator, nutritional content of when swallowed by large animal, 448
Repository for Germinal Choice, 44
Reynolds, R. J. (tobacco company), 35
RIP coefficient, 176
Roberts, Richard, 21, 33, 45, 190
Rocky Mountain Poison Center, 39
Rotation of the earth, effects of sandwich filler on, 79

Safety equipment, recommended, 189
Sagan, Carl, 1
Sakamoto, Junko, 37
San Diego Naval Hospital, 42
Sancetta, Connie, 134
Sandefur, Thomas, 35
Sands, John, 41
Sandvik, Hogne °, 35
Schedule, bus, 428
Schiffman, Jay, 42, 192
Schwartz, Mel, 23
Scientific Gossip, 87, 130, 151, 171, 183
Scientific paper, how to write a, 187
Scratch 'n' Smell demonstration, 98
Sen, Sun-yat, 409
Shannahoff-Khalsa, 37
Shoes, difficulty of finding nice, 373
Siegel, Eric, 157
Singapore, 40
Sleep research, 67
Slogan, inane, 91
Smith, Andrew, 37, 38
Smith, Robert Richard, 66, 130
Social Text (journal), 35, 198
Socks, cleanliness of, 441
Sokal, Alan, 198
Southern Baptist Church of Alabama, 40
Southern Methodist University, 41
Space aliens, probability of abduction by, 85
Spam, 43
Spamer, Earle, 107, 169
Sperm collection, 44
Starling, James, 37
Steiling, Kevin, 42
Stillwell, Thomas, 41
Struchkov, Yuri, 43
Students' attention, method for getting, 177
Students, performance of dead and living compared, 175
Studmuffins of Science Calendar, 170
Superman, 24
Surfer Girl Fungus, 127
Swimsuit sweeties, 169
Symmetra, scientist/supermodel, 33, 97, 168
SymmetraCal, 168

Taddeo, Joseph, 35
Taiwan National Parliament, 37
Talk of the Nation: Science Friday, 31
Taxi driving, 23
Teach or I'll Leave (TOIL) movement, 197
Teachers' guide, 174
Telemarketers, how to demoralize, 17

Telephone numbers, mathematics of, 162
Teller, Edward, 26, 43
Temple University, 42
The, article titles that begin with, 114, 75, 82, 175, 150, 79, 146, 35, 68, 162, 142, 116, 51, 166, 101, 148, 113, 170, 127, 107, 149, 63, 164
Thermometer, pop-up, 148
Thing, indescribably horrible, 362
Time Caplet, 190
Timothy, Tiny, 339
Tisch, Andrew, 35
Toastability, physical limits of, 72
Todpreuss, Lyle, 89
Toilet, reading on the, 24
Toothbrush, big, 149
Tornadoes, 82
Trailer homes, 82
Transmission of gonorrhoea through an inflatable doll °, 46
Trash, 87
Trieste, lunch trucks of °, 384
Turnblade, Donald, 190
2000, Project *AIR*head, 199

Ubiquity, 63
UFO sightings, schedule of, 67
Underwear, wet, 37
University of Buffalo, 35
University of California, San Francisco, 42
Unknown Dentist, tomb of, 149
US Tobacco, 35

Van Daniken, Erich, 44
Vardavas, Stephanie, 134
Veg-O-Matic, 42
Virtual academia, 182
Vos Savant, Marilyn, 191

Wakita, Masumi, 37
Wannabes, Ig Nobel, 45
Watanabe, Shigeru, 37
Watson, James, 16, 122
Welcome, welcome speech, 32
Whacko, Project, 197
Whirligigs, Hungarian ethnomusicology and, 379
Williams, Ted, 66
Woodchucks, 114
Worm, gummy, 178
Wow, variant spellings of, 458

X, Patient, 39, 50

Yadda yadda, socioeconomic predilection for saying, 451
Yagi, E., 43
Yeast, happy, 123
Yeast, transmissibility in marriage, 124
Yeh, Sally, 38
Yew, Singapore Prime Minister Lee Kuan, 40
Yodelling, 66

Zero, 164